D0661028

comap, inc

The UMAP Expository Monograph Series

UMAP Monographs bring the undergraduate student new mathematics and fresh applications of mathematics with no delay between the development of an idea and its implementation in the undergraduate curriculum.

The High Cost of Clean Water:
Models for Water Quality Management
Edward Beltrami, *SUNY at Stony Brook*

Spatial Models of Election Competition
Steven J. Brams, *New York University*

Modeling Tomorrow's Energy System:
Applications of Linear Programming
T. Owen Carroll, *SUNY at Stony Brook*

Elements of the Theory of Generalized Inverses for Matrices
Randall E. Cline, *University of Tennessee*

Introduction to Population Modeling
James C. Frauenthal, *SUNY at Stony Brook*

Smallpox: When Should Routine Vaccination be Discontinued?
James C. Frauenthal, *SUNY at Stony Brook*

Conditional Independence in Applied Probability
Paul E. Pfeiffer, *Rice University*

Topics in the Theory of Voting
Philip D. Straffin, Jr., *Beloit College*

Markov Decision Processes
Paul Thie, *Boston College*

Man in Competition with the Spruce Budworm:
An Application of Differential Equations
Philip M. Tuchinsky, *Ford Motor Company's Research and Engineering Center*

The UMAP Expository Monograph Series

Spatial Models of Election Competition

Steven J. Brams
New York
University

comap, inc.

consortium for mathematics
and its applications, inc.

271 lincoln street, suite no. 4
lexington, ma 02173

Author:

Steven J. Brams
Department of Politics
New York University
New York, NY 10003

This material was prepared with the partial support
of National Science Foundation Grants No. SED76-
19615 A02, SED80-07731, and SPE-8304192.
Recommendations expressed are those of the author
and do not necessarily reflect the views of the NSF
or the copyright holder.

©COMAP, Inc. 1983
ISBN 0-912843-01-2
Printed in USA

Contents

Monograph Description Sheet

Title: SPATIAL MODELS OF ELECTION COMPETITION

Author: Steven J. Brams
          Department of Politics
          New York University
          New York, New York 10003

Intended Audience: This monograph is addressed primarily
    to students in analytically oriented courses in
    political science, secondarily to students in mathe-
    matics courses in which applications and modeling are
    stressed.

Content and level: Except for one optional exercise and
    material in the Appendix, only high-school level mathe-
    matics is assumed.   The emphasis is less on mathematical
    analysis and more on developing an appreciation for
    logical reasoning about elections and substantive prob-
    lems encountered in their analysis.   Some simple one-
    dimensional spatial models are developed, and informal
    results derived and illustrated, using rudimentary mathe-
    matics; a brief, more formal development of two theorems
    using elementary calculus is given in the Appendix.

Political science fields: American government, political
    parties, elections, modern political analysis.

Mathematics fields: Analytic geometry (calculus optional),
    game theory, mathematical modeling, operations research.

Acknowledgements: I wish to thank William H. Riker, Philip
    D. Straffin, Jr., Robert M. Thrall, and two anonymous
    referees for valuable comments on an earlier version of
    this monograph.   Material in this monograph has been
    adapted from Steven J. Brams, The Presidential Election
    Game (New Haven: Yale University Press, 1978), chapters
    1 and 4; and Steven J. Brams and Philip D. Straffin, Jr.,
    "The Entry Problem in a Political Race," (New York
    University Press, 1979).   Neither of these sources con-
    tains exercises and answers.

1. Introduction

The purpose of this monograph is to show how, using only elementary mathematical concepts, the positions political candidates take in an election campaign can be analyzed. Voters are assumed to be distributed along a left-right continuum, and candidates are assumed to take positions along that continuum to maximize their vote totals, given that voters vote for the candidate whose position is closest to theirs.

The analysis begins with two-candidate races, in which the median of the voter distribution is shown to be the optimal, equilibrium position of each candidate. The analysis is then extended to multi-candidate races; possible candidate strategies for different segments of the electorate are explored. The effects of fuzzy candidate positions, and voter indifference and alienation, are also studied.

The basic spatial model is then complicated by assuming that as candidates move toward extremist positions, their utility--as measured by the support they receive from activist voters--increases as their probability of winning simultaneously decreases. Positions that maximize a candidate's expected utility are illustrated. As a final complication, it is shown that if a second issue dimension is introduced, candidate platforms comprising positions on two issues may be subject to a paradox of voting, rendering no candidate position invulnerable to challenges by a competitor.

The substantive focus of the analysis is on presidential elections, with particular attention given first to the problems candidates face in winning their party's nomination in a sequence of state primaries, and then to the problems they face in satisfying different elements within their party that may pull them in different directions in the general election. Throughout the monograph, numerous examples of actual candidate behavior in recent presidential primaries and elections are given to provide an interpretation of the analysis and results.

Before plunging into the analysis, it is fair to ask what benefits logical reasoning and mathematics bring to the study of elections. I will respond in two ways, first with a general statement and then an example.

There is nothing to match the hoopla, pageantry, and excitement of a presidential campaign in American politics. No less dramatic, though quieter, are the strategic, game-like features of a presidential campaign, which often are a good deal more consequential. Given their presence and importance, it seems reasonable to suppose that some tools of modern decision theory and game theory may help to illuminate the competitive character of presidential elections and the strategic interdependence of decisions made at different stages in the campaign.

At a minimum, this approach offers more than good hindsight in trying to determine better and worse strategies in presidential campaigns. For example, consider what good hindsight would say after replaying the "mistakes" of the 1972 campaign: Jimmy Carter should not run for his party's nomination in all states in 1976 because Edmund Muskie had done so in 1972 and lost. Of course, this good hindsight is now bad hindsight, since Carter followed this very strategy and won, which illustrates the dubious scientific status of hindsight.

In contrast to the hindsight approach, I have attempted to develop models that can impart a deeper and more general understanding of underlying factors at work in the presidential election process. By "models" I mean simplified representations that abstract the essential elements of some phenomena or process one wants to study. By deducing consequences from models, one can see more clearly what is happening than one can by trying to deal with reality in all its unmanageable detail. Before beginning this analysis, however, I shall first present some background information on presidential elections.

2. Background

Probably the greatest spectacle in American politics is the quest for the presidency. Campaigns for the presidency may commence a year--or even several years--before the first state caucuses and primaries in a presidential election year as the early entrants lay the groundwork for their campaigns by putting together staffs and sounding out local political leaders and potential contributors. The campaigns of most presidential candidates do not attract wide news coverage, however, until the first caucuses and primaries, which now begin in January (Iowa caucus) and February (New Hampshire primary) of an election year. Then ensues a whirlwind of activity for the next nine months or so that culminates on Election Day in November.

More than half the 50 states today--29 in 1976, plus the District of Columbia--hold primaries from the middle of winter through the late spring of a presidential election year. The remaining states choose delegates to the Democratic and

Republican national party conventions in caucuses in which voters at the local or district level elect delegates to state-wide conventions, who in turn elect delegates to the national party convention. These successive elections of delegates may be carried through two or more stages until national party convention delegates are chosen.

The bewildering variety of rules that govern delegate selection in different caucus states makes it impossible to model a "typical" caucus state. Rules governing the selection of national party convention delegates in primary states also differ considerably, but all primary states share one feature: the voters vote directly for a slate of delegates or the candidates in one election, whereas in caucus states the election occurs in stages and is, therefore, indirect.

To be sure, some primary states, like California, also use caucuses in the preliminary selection of slates of delegates. Moreover, primaries may be open or closed, depending on whether voters can "cross over" and vote for delegates or candidates in the other party contest (open) or must stick to their own party contest (closed). In addition, while the outcomes of most primaries are binding on the delegates, some are only advisory--"beauty contests" is the term that has been coined.

The fact that the primary states include virtually all the large states with the most delegates makes performance in them a critical factor in securing the nomination of one's party. Of course, if no candidate succeeds in gaining a decisive lead over his opponents in the primaries, the locus of decision shifts to the national party convention. But no candidate defeated in the primaries is ever likely to reach this phase, even if he is the incumbent president.[1]

---

[1] Although Lyndon Johnson chose not to run in the Democratic primaries in 1968, Eugene McCarthy's "strong showing" in the New Hampshire primary (while losing with 42 percent of the vote to Johnson's 50 percent write-in vote)--and his expected win in the second primary (Wisconsin)--seem to have been important factors in inducing the incumbent president to withdraw from the 1968 race just prior to the Wisconsin primary.

State primaries, then, are the crucial first phase in a candidate's quest for the presidency. If a candidate, by winning a large proportion of pledged delegates in the primaries, effectively wraps up his party's nomination in this phase, then the party convention provides merely a rubber stamp for the nomination game he has already won.

3. The Primacy of Issues and Their Spatial Representation

I start from the assumption that voters respond to the positions that candidates take on issues in state primaries. This is not to say that nonissue-related factors like personality, ethnicity, religion, or race have no effect on election outcomes but rather that issues take precedence in a voter's decision. Indeed, sometimes these "nonissues" become issues, but for purposes of the subsequent analysis I shall assume issues to be questions of public policy--what the government should and should not do on matters that affect, directly or indirectly, its citizens.

The primacy of issues in presidential elections has now been reasonably well documented over the last ten years.[2] Although most of the research that has been conducted applies to the general election, it would seem even more applicable to primaries, in which party affiliation is not usually a factor. Particularly in states where primaries are closed, with only registered Democrats and registered Republicans eligible to participate in choosing delegates to their respective conventions, it is the issue positions of the candidates running for their party's nomination, not their party identification itself, that assume paramount importance in primaries.[3]

Thus, the rule that excludes nonparty candidates from participating in a party's presidential primary would appear to have a rather important political consequence.[4] It forces voters in a primary election to make choices other than on the basis of party affiliation, which is, of course, the same for all candidates running for their party's nomination.

---

[2]See Key (1966). For a general discussion of the role of issues in presidential elections, see the articles, comments, and rejoinders of Pomper, Boyd, Brody, and Kessel (1972). A more recent assessment can be found in Asher (1976, pp. 86-121, 196-199), and references cited therein; see also Pomper (1975, chap. 8); Nie, Verba, and Petrocik (1976, chaps. 10, 16-18); Niemi and Weisberg (1976, pp. 160-235); and Strong (1977). Still more recently, the significance of issues in a voter's decision has been challenged in Margolis (1977), where it is argued that candidate evaluations and party images--among other factors--still hold important sway; for empirical support, see Kelley and Mirer (1974). This criticism, however, ignores the origins of candidate evaluations and candidate images, which, it seems plausible to assume, ultimately spring from the issue positions of candidates and parties--though perhaps as seen in earlier elections.

[3]Flanigan and Zingale (1975, pp. 130-140). Even in open primary states that permit "crossovers" (14 of 30 in 1976), those voters who cross over from one party to another are probably inclined to do so precisely because of the issue positions of candidates not running in their own party's primary. In 1976, however, issue voting declined in importance. See Miller and Levitin (1976, chap. 7).

[4]Formerly, the winner-take-all feature of voting in primaries was also significant, but now a proportional rule governs the allocation of convention delegates in most primary states. (The main exception in 1976 was the Republican primary in California.)

To be sure, a candidate in a primary may claim that he is the only "true" representative of his party's historical record and ideology. But by making this claim, he is not so much invoking his party label to attract votes as saying that his positions on issues more closely resemble those of his party forebears than the positions of his opponents.

How can the positions of candidates on issues be represented? Start by assuming that there is a single over-riding issue in a campaign on which all candidates must take a definite position. (Later candidates will be allowed to fuzz their positions--and thereby adopt strategies of ambig-uity--as well as take positions on more than one issue.) Assume also that the attitudes of party voters on this issue can be represented along a left-right continuum, which may be interpreted to measure attitudes that range from very liberal (on the left) to very conservative (on the right).[5] I shall not be concerned here with spelling out exactly what "liberal" and "conservative" mean but use this interpretation only to indicate that the attitudes of voters can be scaled along some policy dimension to which the words "liberal" and "con-servative" can in some way be meaningfully attached.

I assume that the positions candidates take on this dimension or issue are perceived by voters in the same way-- that is, there is no misinformation about where on the continuum each candidate stands. Like all theoretical assump-tions used to model empirical phenomena, this assumption simplifies the reality of the positions candidates take, and their perceptions by voters, but it serves as a useful starting point for the analysis.

To derive the behavior of voters from their attitudes and the positions candidates take in a campaign, some

---

[5]An issue on which attitudes can be indexed by some quantitative variable, like "degree of government intervention in the economy," obvi-ously better satisfies this assumption than an issue that poses an either-or question--for example, whether or not to support the development of a major new weapons system.

assumption is necessary about how voters decide for whom to vote. I am not concerned with the attitudes of <u>individual</u> voters, however, but only with the <u>numbers</u> who have particular attitudes along some liberal-conservative scale.

For this purpose I postulate a <u>distribution</u> of voters, as shown in Figure 1. The vertical height of this distribution, which is defined by the curve in Figure 1, represents

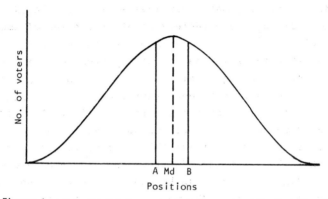

Figure 1. Two candidates: symmetric, unimodal distribution.

the number (or percentage) of voters who have attitudes at each point along the horizontal continuum.[6]

Because the distribution I have postulated has one peak, or mode, it is characterized as <u>unimodal</u>. Since the curve has the same shape to the left and the right of its median, which

---

[6]This spatial representation of voter attitudes and candidate positions was first used in Downs (1957). For a critical assessment of this work, see Stokes (1963); Robertson (1976), which tests predictions of the theory for the British electorate; and Frohlich, Oppenheimer, Smith, and Young (1978). For a review of the more recent literature on party-competition models, see Riker and Ordeshook (1973, chaps. 11 and 12); Shepsle (1974, pp. 4-77); Taylor (1975, pp. 413-481); and Ordeshook (1976, pp. 285-313).

is the point where the vertical dashed line intersects the horizontal axis, the distribution is <u>symmetric</u>.[7]

I have also postulated in Figure 1 the positions of two candidates, A and B, at points along the left-right continuum. Assume that candidate A takes a position somewhere to the left of the median and candidate B a position somewhere to the right. How attractive are these positions to the voters? This is the question I turn to in section 4, where the analysis is restricted to competition between just two candidates; in section 5, I shall consider what happens when more than two candidates enter the race.

---

[7]A <u>median</u> divides the area under a distribution curve exactly in half, which means in our example that half the voters have attitudes to the left of the point where the median line intersects the horizontal axis and half the voters have attitudes to the right of this point. Moreover, because the distribution is symmetric--the curve to the left of the median is a mirror image of the curve to the right--the same numbers of voters have attitudes equal distances to the left and right of the median.

4. Rational Positions
in aTwo-Candidate Race

I assume that both voters and candidates have goals in an election, and they act rationally to satisfy these goals. To act rationally means simply to choose the course of action that best satisfies one's goals.

The rationality assumption is rather empty unless particular goals are postulated for voters and candidates. For voters, I assume that they will vote for the candidate whose position is closest to their own along the continuum. For candidates, I assume that they will try to choose positions that maximize the total number of votes they receive, in light of the voters' rationality.[8]

---

[8]Alternative models in which candidates have policy preferences and view winning as a means to implement them--rather than more cynically adopt policy positions as a means to winning--are developed in Wittman (1973); Wittman (1977); and Wittman (1976); see also McKelvey (1975). Policy considerations, based on the assumption that utilities are associated with different candidate positions, will be introduced in a model in section 9.

While the <u>attitudes</u> of voters are a fixed quantity in the calculations of candidates, the <u>decisions</u> of voters will depend on the positions the candidates take. Given the candidates know the distribution of voter attitudes, what positions for them are rational?

Assume that there are only two candidates in the race, and the distribution of voters is symmetric and unimodal, as illustrated in Figure 1. If candidates A and B take the positions shown in Figure 1, A will certainly attract all the voters to the left of his position, and B all the voters to the right of his position. If both candidates are an equal distance from the median, they will split the vote in the middle (the left half going to A and the right half going to B). The race will therefore end in a tie, with half the votes (to the left of the median) going to A and half the votes (to the right of the median) going to B.

Could either candidate do better by changing his position? If B's position remains fixed, A could move alongside B, just to his left, and capture all the votes to B's left. Since A would have moved to the right of the median, he would, by changing his position in this manner, receive a majority of the votes and thereby win the election.[9]

But, using an analogous argument for B, there is no rational reason for him to stick to his original position to the right of the median. He should approach A's original position to capture more votes to his right. In other words, both candidates, acting rationally, should approach each other and the median. Should one candidate (say, A) move past the median, but the other (B) stop at the median, B

---

[9] I assume for now that A does not suffer any electoral penalty at the polls from changing his position, though fluctuations along the continuum may evoke a charge of being "wishy-washy," which is a feature of candidate positions that I shall analyze in section 8. Alternatively, the "movements" discussed here may be thought to occur mostly in the minds of the candidates before they announce their actual positions.

would receive not only the 50 percent of the votes to his left but also some votes to his right that fall between his (median) position and A's position (now to B's right). Hence, there is not only an incentive for both candidates to move toward the median but not to overstep it as well.

The consequence of these calculations is that the median position is optimal for both candidates. Presumably, if they both adopted the median position, voters would be indifferent to the choice between the two candidates on the basis of their positions alone and would make their choice on some other grounds.

More formally, the median position is optimal for a candidate if there is no other position that can guarantee him a better outcome (i.e., more votes), regardless of what position the other candidate adopts. Naturally, if B adopted the position shown for him in Figure 1, it would be rational for A to move alongside him to maximize his vote total, as I have already demonstrated. But this nonmedian position of A would not ensure him of 50 percent of the votes if B did not remain fixed but instead switched his position (say, to the median). Thus, the median is optimal in our example in the sense that it guarantees a candidate at least 50 percent of the total vote no matter what the other candidate does.

Exercise 1. Define a candidate's position in a two-candidate race to be opposition-optimal if, given the position of an opponent is fixed, it maximizes his (the first candidate's) vote total. Show that a candidate's opposition-optimal position must be adjacent to his opponent's position. (Roughly speaking, "adjacency" means an infinitesimal distance away.)

Exercise 2. If the fixed position of an opponent in a two-candidate race is not at the median, show that a candidate's opposition-optimal position is adjacent to his opponent's and closer to the median.

The median is also "stable" in our example because, if one candidate adopts this position, the other candidate has no incentive to choose any other position. More formally, a position is in equilibrium if, given it is chosen by both candidates, neither candidate is motivated unilaterally to depart from it. Thus, the median in our example is both optimal (offers a guarantee of a minimum number of votes) and is in equilibrium (once chosen by both candidates, there is no incentive for either unilaterally to depart from it).

A surprising consequence of all two-candidate elections is that, whatever the distribution of attitudes among the electorate, the median loses none of its appeal in a single-issue election. Consider the distribution of the electorate in Figure 2, which is bimodal (i.e., has two peaks) and is

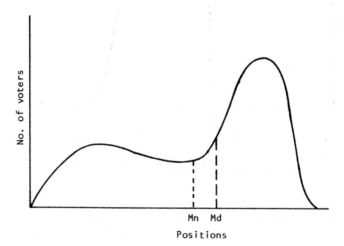

Figure 2. Nonsymmetric, bimodal distribution in which median and mean do not coincide.

not symmetric. Applying the logic of our previous analysis, it is not difficult to show that the median is once again the optimal, equilibrium position for two candidates.

In this case, however, the mean (Mn), which is the point at which the voters, weighted by their positions along the continuum, are balanced on the left and right of Mn, does not coincide with the median. This is because the distribution is skewed to the right, which necessarily pushes the median to the right of the mean. A sufficient condition for the median and mean to coincide is that the distribution be symmetric, but this condition is not necessary: the median and mean may still coincide if a distribution is nonsymmetric, as illustrated in Figure 3.

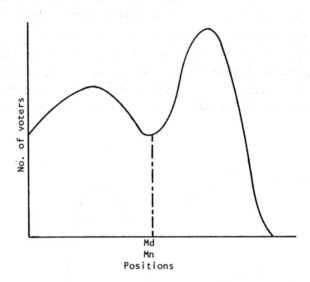

Figure 3.   Nonsymmetric, bimodal distribution in which mean and median coincide.

The lesson derived from Figure 2 is that it may not be rational for a candidate to take a "weighted average" position on an issue (i.e., at the mean) if the distribution of attitudes of the electorate is skewed to the left or right. Figure 3 indicates, however, that the noncoincidence of the median and mean is not necessarily related to the lack of symmetry in a distribution: half the voters may still lie to the left, and half to the right, of the mean (as well as the median) if the distribution is nonsymmetric.

Exercise 3. As a rough approximation to the continuous distribution in Figure 2, consider the following discrete distribution of 19 voters whose positions on a 0-1 scale are as follows:

1 voter at 0.1
3 voters at 0.2
2 voters at 0.3
2 voters at 0.5
3 voters at 0.6
6 voters at 0.8
2 voters at 0.9

What is the median position? What is the mean?

Exercise 4. As a rough approximation to the continuous distribution in Figure 3, consider the following discrete distribution of 25 voters whose positions on a 0-1 scale are as follows:

2 voters at 0.0
3 voters at 0.2
4 voters at 0.3
3 voters at 0.4
2 voters at 0.5
4 voters at 0.7
6 voters at 0.8
1 voter at   0.9

What is the median position? What is the mean?

Given the desirability of the median position in a two-candiate, single-issue election, is it any wonder why candidates who prize winning try so hard to avoid extreme positions? Even, as in Figures 2 and 3, when the greatest concentration of voters does not lie at the median but instead at a mode (the mode to the right of the median in both these figures), a candidate would be foolish to adopt this modal position. For although he may very much please right-leaning voters, his opponent, by sidling up to this position but still staying to the left of the mode, would win the votes of a majority of voters.

Voters on the far left may not be particularly pleased to see both candidates situate themselves at or near the right-hand modes in Figures 2 and 3, but in a two-person race they have nobody else to whom to turn. Of course, if left-leaning voters should feel sufficiently alienated by both candidates, they may decide not to vote at all, which has implications for the analysis that will be explored in section 8.

I conclude this section by mentioning a rather different application of the analysis as it has been developed so far. This application is to business, which in fact was the first substantive area to which spatial analysis was applied.[10] Consider two competitive retail businesses (say, department stores) that consider locating their stores somewhere along the main street that runs through a city. Assume that, because transportation is costly, people will buy at the department store nearest to them. Then the analysis says that, however the population is distributed along (or near) the main street, the best location is the median. If the city's population is uniformly distributed (i.e., not concen-trated at one end or the other of the main street), then this location will, of course, be at the center of the main street.

Indeed, clusters of similar stores are frequently bunched together near the center of the main street, though these stores may not be particularly convenient to people who live far from the city's center (i.e., median/mean, if the city's population is uniformly distributed)--and, consequently, not in the public interest since their location discriminates

---

[10]See Hotelling (1929); Lerner and Singer (1937); and Smithies (1941).

against these people.[11] To accommodate shoppers in the suburbs as their density has increased over the years, however, shopping centers have sprung up, which--in terms of the previous analysis--says that new candidates have been motivated to enter the race.

The rationality of entry into a political race is an interesting but almost totally neglected question in the study of elections. Because presidential primaries, especially at the start of the sequence, tend to attract many candidates, it seems useful to ask what conditions make entry in a multi-candidate race attractive.

---

[11]Hotelling (1929, p. 53). The social optimum, Hotelling argues, would be for the stores to locate at the 1/4 and 3/4 points along the main street so that no customer would have to travel more than 1/4 of the length of the street to buy at one store. On the other hand, one might argue that if both stores were located at the center, the public interest would be served because greater competition would be fostered.

5. Rational Positions in a Multi-Candidate Race

If there are no positions that a potential candidate can take in a primary that offer some possibility of success, then it will not be rational for him to enter the race in the first place. For a potential candidate, then, the rationality of entering a race, and the rationality of the positions he might take once he enters, really pose the same question.

Assume that two candidates have already entered a primary, and consistent with the analysis in section 4, they both take the median position (or positions very close to it so that they are effectively indistinguishable). Is there any "room" for a third candidate?[12]

---

[12]This question is considered briefly in Robertson (1976, Appendix 1) in the context of an electorate that changes with the enfranchisement of new voters. In light of the subsequent analysis, Robertson's statement that "all that we say [about a two-party system] can be generalised to multiparty systems without too much difficulty" (p. 7) is hard to accept.

Consider Figure 1, but now imagine that A and B have both moved to the median and therefore split the vote since they take the same position. Now if a third candidate C enters and takes a position on either side of the median (say, to the right), it is easy to demonstrate that the area under the distribution to C's right may encompass <u>less than</u> 1/3 of the total area under the distribution curve and still enable C to win a plurality of votes.

To see why this is so, in Figure 4 I have designated, for a position of C to the right of A/B (at the median), the

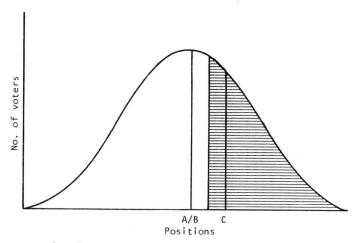

Figure 4. Three candidates: symmetric, unimodal distribution.

portion of the electorate's votes that A/B on the one hand, and C, on the other, would receive. If C's area (shaded) is greater than 1/2 of A/B's area (unshaded), he will win more votes than A or B. (Recall that A and B split their portion of the vote since they take the same [median] position.)

Now C's area includes not only the voters to the right of his position but also some voters to his left. More pre-cisely, he will attract voters up to the point midway between his position on the horizontal axis and that of A/B: A

- 21 -

and B will split the votes to the left of this point, C will win all the votes to the right of this point. Since C picks up some votes to the left of his position, this is why less than 1/3 of the electorate can lie at or to his right and he can still win a plurality of more than 1/3 of the total vote.

Exercise 5. For the voter distribution given in Exercise 4, assume C's position is at 0.8 and A/B's at the median. Verify that the proposition that "less than 1/3 of the electorate can lie at or to his [C's] right and he [C] can still win a plurality of more than 1/3 of the total vote" is true.

By similar reasoning, it is possible to show that a fourth candidate D could take a position to the left of A/B and further chip away at the total of the two centrists. Indeed, D could beat candidate C as well as A and B if he moved closer to A/B (from the left) than C moved (from the right).

Clearly, the median position has little appeal, and is in fact quite vulnerable, to a third or fourth candidate contemplating a run against two centrists. This is one lesson that centrist candidates Hubert Humphrey and Edmund Muskie learned to their dismay in the early Democratic primaries in 1972 when George McGovern and George Wallace mounted challenges from the left and right, respectively. Only after Muskie was eliminated, and Wallace was disabled by an assassin and forced to withdraw, did Humphrey begin to make gains on McGovern in the later primaries, but not by enough to win.

In fact, there are no positions in a two-candidate race, for practically any distribution of the electorate, in which at least one of the two candidates cannot be beaten by a third (or fourth) candidate.[13] I have already shown that both candidates in a two-candidate race can be beaten by a third (or fourth) candidate if they both adopt the median position. Indeed, it is easy to show that whatever position two candidates adopt (not necessarily the same), one will always be vulnerable to a third candidate; if the other is not, he will be vulnerable to a fourth candidate.[14]

What if two candidates, perhaps anticipating other entrants and realizing the vulnerability of the median, take different positions, as illustrated in Figure 5? In this example, because the distribution is bimodal (as well as being symmetric), positions at the modes would seem strong positions for each of two candidates to hold.

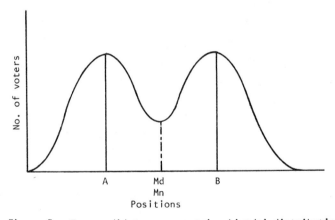

Figure 5. Two candidates: symmetric, bimodal distribution.

---

[13]Given certain assumptions, there are equilibrium positions as the number of candidates increases and the original candidates are free to change their positions, too, but this fact does not inhibit the entry of new candidates (see note 16 below). Lerner and Singer (1937, pp. 176-182) provide details on equilibria in multi-candidate races, though their analysis is developed for buyers and sellers in a competitive market.

[14]For details, see the Appendix.

But enter now a right-leaning third candidate C, who would like to push candidate B out of the race. Excluding the possibility of ties, either there are (i) more voters to the right of B than between B and the median/mean or (ii) the opposite is true. If (i) is true, then C can beat B by moving alongside B to his right; if (ii) is true, then C can beat B by moving alongside B to his left. In either event, B is vulnerable to a third candidate C (and A would be vulnerable to a fourth candidate D for similar reasons). Hence, a third (or fourth) candidate can, by himself, knock out at least one of the two original candidates (A and B) in our example.

Exercise 6. Define a joint defensive-optimal strategy of two candidates to be one which makes it impossible for a third candidate to defeat both of them. (As was shown in the text, it is always possible for a third candidate to defeat at least one of the two original candidates, whatever their positions.) Can you think of a joint defensive-optimal strategy of two candidates--that is, one that would prevent the defeat of one of them by a third candidate?

Exercise 7. Does it seem plausible that two candidates would consciously plan their electoral strategies together to make entry by a third candidate unrewarding? Does a fortuitous choice of such strategies by the two original candidates seem plausible?

Exercise 8. Can you think of a joint defensive-optimal strategy of two candidates that would prevent defeat of one of them by a third and fourth candidate?

It is clear from the answers to the preceding exercises that any positions that two candidates might take in single-issue races are vulnerable to third and fourth candidates. There is, in fact, always a place along a left-right continuum at which a new candidate can locate himself that will displace one or more nearby candidates.

This conclusion is in direct conflict with Anthony Downs's assertion that "there is a limit to the number of parties [candidates in the present analysis] which can be supported by any one distribution. When that limit is reached, no more parties can be successfully introduced."[15] On the contrary, no such limit exists, for reasons already given.[16]

This analysis thus provides an explanation, in terms of the rational choices of both voters and candidates, why many candidates may initially be drawn into the primary fray. As cases in point, in the first Democratic primary in New Hampshire in 1976, four candidates each received more than 10 percent of the vote, while in the second primary in Massachusetts seven candidates each received at least 5 percent of the vote. In neither primary did the front-runner (Jimmy Carter in New Hampshire, Henry Jackson in Massachusetts) receive as much as 30 percent of the total Democratic vote.

---

[15]Downs (1957, p. 123).

[16]Downs seems falsely to have thought that (i) his assumption that a party is not perfectly mobile--"cannot leap over the heads of its neighbors" once it has come into being--would prevent disequilibrium; (ii) once equilibrium is reached, "new parties . . . cannot upset" it (Downs, 1957, p. 123). With respect to (i), a form of cooperation--not just competition with restricted mobility--that allows the parties to make simultaneous adjustments seems also necessary for parties to reach equilibrium positions (assuming they exist); with respect to (ii), the concept of equilibrium implies only that no old party can benefit from unilaterally shifting its position but says nothing about the benefits-- discussed in the text--that may accrue to new parties that take up other positions along the continuum.

6. The Winnowing-Out Process in Primaries

So far I have restricted the spatial analysis of presidential primaries to a single election in which the positions that candidates take on a single issue totally determine the vote they receive. Unlike the general election, however, in which the party affiliation of a presidential candidate may account for a substantial portion of his vote independent of the position he takes on any issue, the assumption that a candidate's position on an issue is determinative does not seem an unreasonable one from which to launch an analysis of primaries. Indeed, most candidates in presidential primaries tend to be identified as "liberal," "moderate," or "conservative," based on their positions on a range of domestic and foreign policy questions. (In section 13, however, I shall show that if there are multiple issues on which candidates are simultaneously evaluated, the simple one-dimensional spatial analysis heretofore described may not yield optimal positions that are in equilibrium.)

The spatial analysis in section 5 suggested why many candidates are drawn into the presidential primaries. To be sure, if an incumbent president or vice president is running, or even contemplates running, members of his party may be deterred from entering the primaries because of the built-in advantages that his incumbency brings.[17] But, it should be pointed out, incumbency did not stop Eugene McCarthy from challenging Lyndon Johnson in the 1968 Democratic primaries, Paul McCloskey from challenging Richard Nixon in the 1972 Republican primaries, or Ronald Reagan from challenging Gerald Ford in the 1976 Republican primaries.

Generally speaking, most primary challenges that have been mounted against an incumbent in recent presidential elections have been single-man crusades and can be viewed, therefore, as essentially two-candidate contests. On the other hand, when an incumbent does not run, the field opens up and many candidates are motivated to stake out claims at various points along the left-right continuum, as I showed earlier.

To explain the entry of multiple candidates into primaries, I considered the contest for the nomination as if it were one election in which each candidate sought to maximize his vote total. But this limited perspective clearly will not do to explain the exit of candidates from primaries. Indeed, probably the most important feature of presidential primaries distinguishing them from other elections is their sequential nature; it is performance in the sequence--not in one primary election--that is crucial to a candidate's success.

This fact is conveyed quite dramatically by statistics from the 1972 Democratic primaries. In these primaries, roughly 16 million votes were cast, with George McGovern

---

[17]For a rational analysis of this question, see Brams (1976, pp. 126-135), and references cited therein.

polling 25.3 percent of the total primary vote and Hubert
Humphrey 25.4 percent, despite entering late.[18] Nonetheless,
though McGovern received fewer primary votes than Humphrey,
and little more than a quarter of the total, he went on to
win his party's nomination on the first ballot at the
national convention.

Hugh A. Bone and Austin Ranney attribute McGovern's
success "to certain breaks,"[19] but it seems that a winning
strategy in a series of primaries is more than a matter of
luck. I shall not try to analyze McGovern's success specifically, however, but rather attempt to identify optimal
strategies over a sequence of elections generally.

As an institution, one is immediately struck by the
fact that primaries play less of a role in selecting candidates than in eliminating them. Candidates who have won or
done well in the primaries, such as Estes Kefauver in the 1952
Democratic primaries or Eugene McCarthy in the 1968 Democratic
primaries, have, despite their impressive showings, lost their
party's nomination to candidates who did not enter the primaries (Adlai Stevenson in 1952, Hubert Humphrey in 1968).
No candidate who has been defeated in the primaries, however,
has ever gone on to capture his party's nomination in the
convention.

Once a candidate enters the primaries, his first-priority
goal is not to be eliminated. In a multi-candidate race, this
goal most often translates into not being defeated by an
opponent, or opponents, who appeal to the same segment of the
party electorate.

For convenience, assume that there are three identifiable
segments of the party electorate: liberal, moderate, and
conservative. This trichotomization of the electorate may

---

[18]Bone and Ranney (1976, p. 81).
[19]Bone and Ranney (1976, p. 81).

not always be an accurate way of categorizing different positions in multi-candidate races, but these labels are commonly used by the media and the public.

A candidate who takes a position on the left-right continuum will, I assume, fall into one of these three segments. Depending on the segment he is identified with, he will be viewed to be in a contest--at least in the first primaries--with only those other candidates who take positions in this segment.

What is likely to happen if there are at least three candidates contesting the vote in each segment? More specifically, who is likely to beat whom in the first-round battles and survive the cuts of candidates in each segment?

If the distribution of the electorate is symmetric and unimodal, as pictured in Figure 1, then the liberal segment will appear as in Figure 6, with the median of this segment to the right of the mean. For reasons given in section 4, the

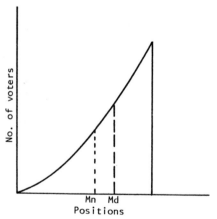

Figure 6. Liberal segment of symmetric, unimodal distribution.

median will be attractive in a two-candidate liberal contest, but should a third candidate battle two candidates who take the median position in this segment, then his rational strategy

-29-

would be to move to the right of the median--and toward the center of the overall distribution--where more of the voters are concentrated in the liberal, and adjoining moderate, segments.

This movement toward the center may be reinforced by two considerations, one related to the concentration of votes near the center and the other by an anticipation of future possibilities in the race. As discussed in section 8, if voters become alienated by a candidate whose position is too far from their own, and respond by not voting, a candidate would minimize this problem by being to the right rather than the left of the median in Figure 6, where a loss a given distance from his position would be numerically less damaging. In addition, a position to the right of the median is more attractive as moderate candidates are eliminated and the liberal survivor can begin to encroach on voters who fall into the moderate segment.

Thus, liberal candidates will be motivated to move toward the moderate segment and, for analogous reasons, conservative candidates will also be motivated to move toward the moderate segment (though from the opposite direction). What should the moderates do in their own segment (see Figure 7)?

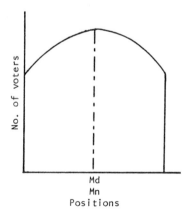

Figure 7. Moderate segment of symmetric, unimodal distribution.

If two candidates take the median position, which is
also the mean because of the symmetry of this segment, then
a third moderate candidate would be indifferent to taking a
position to the left or right of the median/mean since voters
are symmetrically distributed on either side. To illustrate
the consequences of a nonmedian position, assume that the
third candidate takes a position somewhat to the right in the
moderate segment. He thereby captures a plurality of the
moderate votes against his two opponents at the median (for
reasons given in section 5 for the entire distribution) and
eliminates them from the contest.

If, as I argued earlier, a moderate-leaning liberal and
a moderate-leaning conservative are advantaged in their segments in multi-candidate contests, they can eliminate their
median opponents from the respective contests on the left and
right. As a consequence of these outcomes, the election
would reduce to a three-way contest among a liberal (L), a
moderate (M), and a conservative (C), with positions approxi-
mately as shown in Figure 8. (As indicated earlier, I assume

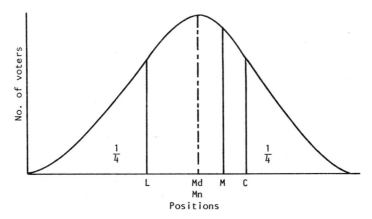

Figure 8.   Three-way contest among liberal, moderate, and
conservative candidates.

that the moderate takes a position to the right of the median/
mean.)

-31-

In this manner, the initial primaries serve the purpose of reducing the serious candidates in each segment to just one. But the elimination process does not stop here. In fact, if as few as 1/4 of the voters lie to the left, and 1/4 of the voters lie to the right, of the liberal and conservative candidates, respectively (see Figure 8), is is unlikely that the moderate candidate will get the most votes. For, by the previous assumption, he is not at the median but to its right, so he will in all likelihood receive hardly more than 1/2 of those votes in the middle (or 1/4 of the total, since 1/2 of the total fall between L and C).[20]

Hence, the moderate candidate will probably receive fewer votes than the liberal candidate and perhaps fewer than the conservative candidate as well. For both the liberal and conservative candidates will pick up all the votes to their left and right, respectively (1/4 of the total), plus all votes in the moderate segment up to the point midway between their positions and those of the moderate candidate. In fact, if the liberal and conservative candidates can supplement their 1/4 liberal and 1/4 conservative support with as few as an additional 1/12 of the total votes from the moderate segment, they would each receive 1/3 of the total and thereby limit the moderate candidate to 1/3, too.

---

[20]If the moderate candidate's position were at the median, he would receive more than 1/2 the votes between the points L and C since voters are more concentrated around the median than at L or C. But being to the right of the median, the votes that would be divided between him and the liberal candidate at the point midway between L and M would, if he were sufficiently far away from Md/Mn, give the advantage to the liberal candidate. The conservative candidate would get fewer votes the closer the moderate candidate approached him, but, depending on the distribution, it is certainly possible that the liberal and conservative could both beat the moderate in the three-way contest depicted in Figure 8.

Exercise 9. As a rough approximation to the continuous distribution in Figure 8, consider the following discrete distribution of 25 voters whose positions on a 0-1 scale are as follows:

1 voter at 0.1
2 voters at 0.2
3 voters at 0.3
4 voters at 0.4
5 voters at 0.5
4 voters at 0.6
3 voters at 0.7
2 voters at 0.8
1 voter at 0.9

Assume L is at position 0.3 (6 voters, or 24 percent, at or to his left) and C is at position 0.7 (6 voters, or 24 percent, at or to his right). If M is at 0.6 (slightly to the right of Md = 0.5, as indicated in Figure 8), would L and C succeed in limiting him to less than 1/3 of the total vote? How would L and C do?

Exercise 10. Is there any position that M can take between L and C that would guarantee him victory in the election?

Because of the vulnerability of the center to simultaneous challenges from the left and right, it is really not surprising that a liberal candidate like McGovern could win his party's nomination with only slightly more than 25 percent of the primary votes. More generally, a moderate candidate can be squeezed out of the race by challengers on both sides of the spectrum even when the bulk of voters fall in the middle. If most voters are not concentrated in the middle, but tend instead to be either liberal or conservative, then of course the problems of a moderate are aggravated.

Exercise 11. For the bimodal voter distribution given in Exercise 4, show that there is no position between L at 0.3 and C at 0.7 that would result in M's receiving more than 5 votes, or 20 percent of the total.

Even if most voters are concentrated in the middle, the moderate may face another kind of problem. Contrary to the model postulated earlier, more than one moderate may attract a sufficient number of votes to survive the early primaries. But opposed by just one surviving liberal and one surviving conservative in the later primaries, the two or more moderates who divide the centrist vote will lose votes as the primaries proceed, relative to the liberal and conservative candidates who pick up votes from those in their segment whom they eliminate. The 1964 Republican primaries are an example of this situation, in which Henry Cabot Lodge, Jr., a moderate, lost out to Nelson Rockefeller and Barry Goldwater, the liberal and conservative candidates who fought a final climactic battle in the Califormia primary that Goldwater won.

Moderates are not inevitably displaced in a sequence of primaries--as the case of Jimmy Carter in the 1976 Democratic primaries demonstrates--but this has been one trend in recent years in heavily contested primaries in both parties. As I have tried to show, spatial analysis enables one to understand quite well the weakness of moderates when squeezed from the left and right in a series of elimination contests.

7. The Factor of Timing

Primaries, I have suggested, are first and foremost elimination contests that pare down the field of contenders over time. Implicit in the previous analysis has been the assumption that the key to victory in the primaries is the position that a candidate takes on a left-right continuum in relation to the positions taken by other candidates. Thus, a candidate's goal of avoiding elimination, and eventually winning, cannot be pursued independently of the strategies other candidates follow in pursuit of the same goal. This quality of primaries, and elections generally, is what gives such contests the characteristics of a game, in which winning depends on the choices that all players make.

Since the rules of primaries do not prescribe that these choices be simultaneous,[21] there would appear to be advantages in choosing after the other players have committed themselves and the strengths and weaknesses of their positions can be better assessed. Indeed, some candidates avoid the early primaries, and join the fray at a later stage, on the basis of just such strategic calculations. Robert Kennedy, for example, stayed out of the 1968 Democratic primaries until the weakness of Lyndon Johnson's position as the incumbent became apparent, and Johnson had withdrawn from the race, before engaging Eugene McCarthy in Indiana and the later primaries.

A more extreme case of a late-starter was Hubert Humphrey, who stayed out of the 1968 Democratic primaries altogether, apparently believing that as the incumbent vice president he stood his best chance in the national party convention. He was not to be disappointed, winning on the first ballot in the convention, though his only serious opposition came from McCarthy because of the earlier assassination of Kennedy after the California primary.

The advantages of starting late, when the positions of one's opponents are known and their weaknesses can be identified and exploited, must be balanced against the organizational difficulties one faces in launching a campaign hurriedly. Last-minute efforts by even well-known candidates have often fizzled out.

The campaigns of some late-starters do take off, however, as illustrated by Robert Kennedy's run for the 1968 Democratic

---

[21] In some states, these choices are not made by the candidates at all but by a state official who places the names of all recognized candidates on the ballot, whether they have formally announced their candidacies or not. In other states, there are filing dates that must be met if one's name is to appear on the ballot. But even these can be ignored in most states if one runs as a write-in candidate. However, successful write-in campaigns, especially by nonincumbents, are rare, notwithstanding Henry Cabot Lodge, Jr.'s victory as a write-in in the 1964 Republican primary in New Hampshire.

nomination before he was assassinated. True, it is usually
only already well-known contenders who enjoy the privilege
of holding out on announcing their candidacies. Candidates
who came from nowhere, like Eugene McCarthy in 1968, George
McGovern in 1972, and Jimmy Carter in 1976, have no choice
but to start their campaigns very early in order to acquire
sufficient recognition to make a serious run.

How can spatial analysis be used to model the factor
of timing? Consider the situation in which several candidates
to the left and right of the median struggle for their party's
nomination in the early primaries. Assume that their various
positions fall within the shaded bands pictured in Figure 9,
in which the distribution of voter attitudes is assumed to be
symmetric and unimodal.

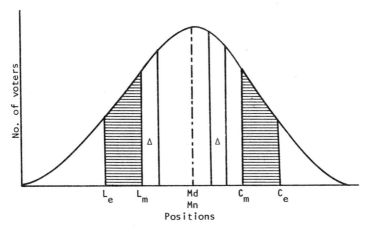

Figure 9. Bands encompassing positions of candidates on left
and right.

Assume that a prominent moderate politician considers
making a bid for his party's nomination by positioning himself
somewhere near the median/mean. He calculates that his
chances of winning his party's nomination are good if extreme
(e) candidates are the ones to survive in the early primaries
on the left and right (at positions L_e and C_e), since he will

be able to capture the bulk of the votes in the middle of the distribution. On the other hand, if moderate (m) candidates are the ones to survive in the early primaries (at positions L_m and C_m), he will probably be squeezed out by one or the other if he runs, for reasons given in section 6.

Thus, to gain a better picture of his chances, the prominent moderate may decide to await the results of the early primaries before making his decision, even if it means postponing the building of a campaign organization that would enable him to make a stronger bid. Aside from the problem of organizing an effective campaign late in the game, however, there may be a more compelling reason to avoid an announcement, based on spatial considerations.

Assume that the survivors of the early primaries are an extreme liberal candidate (at L_e) and a moderate conservative candidate (at C_m). Thus, if the moderate runs, he would be squeezed more from the right than from the left. Clearly, his chances are not so favorable as they would be if he faced two extreme candidates on the left and right. Nonetheless, what spatial analysis clarifies is how he can capitalize on the information he gains from awaiting the results of the early primaries to position himself optimally against his two surviving opponents at L_e and C_m.

Although one might think initially that a hold-out moderate could maximize his vote total by taking a position midway between L_e and C_m, a glance at Figure 9 will show this to be a poor strategy. Instead, he should take a position to the right of the median/mean near C_m.

The latter strategy follows from the fact that the votes he gives up to his L_e opponent as he moves to the right of the median/mean are more than compensated for by the votes he gains from his C_m opponent as he moves toward his position. Visually, it can be seen from Figure 9 that there are more votes in the Δ-region just past the midway point between the median/mean and C_m than in the Δ-region just past the midway point between L_e and the median/mean. Therefore, a moderate gains more votes (in the right Δ-region) than he loses (in the left Δ-region) as he moves rightward toward C_m.

We see, then, that if the distribution of voter attitudes is symmetric and unimodal, a late-starting moderate's best weapon against opponents on his left and right is to move toward his more moderate opponent. Our qualitative analysis does not say exactly how far he should move, but this is a problem that can easily be solved if the distribution of voter attitudes is known.

Exercise 12. For the symmetric, unimodal voter distribution given in Exercise 9, assume L_e is at 0.2 and C_m is at 0.7. Show that the position m of a moderate M that maximizes his vote total is not at the median 0.5 but to the right of Md.

Exercise 13 (optional). Consider the continuous density function $f(x) = 6(x-x^2)$, which defines the (unique) parabola, symmetrical about a vertical axis, that passes through points $(0,0)$ and $(1,0)$ and whose area in the interval $0 \leq x \leq 1$ is

$$\int_0^1 6(x-x^2)dx = 1.$$

As in Exercise 12, assume that L_e is at 0.2 and C_m is at 0.7. Draw a graph of the voter distribution curve defined by $f(x)$ and show that the position of a moderate M that maximizes his vote total is m = 0.55.

The analysis in this section can be extended to different-shaped distributions and can incorporate different assumptions about the positions of committed candidates and the timing of the announcement of an uncommitted candidate. My main purpose, however, has been to introduce with a simple example the factor of timing into the spatial analysis of primaries, not to try to treat this subject exhaustively. It is a subject that deserves much more systematic attention than it has received in the literature.

8. Fuzzy Positions and Alienation

In section 7 I considered the possibility that there may be several candidates to the left of the median, and several candidates to the right, whose collective positions can be represented by bands, rather than lines, on the distribution. This same representation can also be used to model the positions of candidates that are fuzzy, i.e., that cover a range on the left-right continuum instead of occurring at a single point on the continuum.

Fuzzy positions in campaigns are well-known and reflected in such statements as, "I will give careful consideration to . . ." (all positions are open and presumably equally likely), "I am leaning toward . . ." (one position is favored over the others but not a certain choice), and "I will do this if such and such . . ." (choices depend on such-and-such factors). Such ambiguous statements may be interpreted as probability distributions, or lotteries, over specific positions and have been shown, under certain circumstances, to be rational choices not only for candidates but for voters as well.[22]

To model fuzzy positions, I shall not introduce proba-
bilities into the spatial analysis but instead shall analyze
some implications of band versus point positions. First,
however, to motivate the subsequent analysis, consider why
a candidate may not want to adopt a clear-cut position on an
issue.

Perhaps the principal disadvantage of clarity in a
campaign is that, while attracting some voters, it may
alienate others, independently of the positions that other
candidates take. That is, voters sufficiently far from the
position that a candidate takes at a particular point on the
continuum may feel disaffected enough not to vote at all,
even given the fact that his position is closer to theirs
than that of any other candidate.

Much has been made of the "alienated voter" in the voting
behavior literature, with many different reasons offered for
his alienation.[23] Although there is not universal agreement
on why voters are alienated, the fact of alienation--as
measured, for example, by the number of citizens who fail to
vote--is indisputable. To be sure, some voters fail to vote
because of legal restrictions (e.g., residency requirements),
but the vast majority of nonvoters in a presidential
election--an average of about 40 percent in recent presidential elections,[24] which climbed to a historic high of 46
percent in 1976[25]--are eligible but choose not to exercise
their franchise. In competitive primaries, by comparison,
an even greater proportion of eligible voters--an average of

---

[22]See Brams (1976, pp. 53-65), and references cited therein.

[23]The classic study is Levin (1960). For recent analyses, see
Wright (1976); and several articles on "Political Alienation in America"
(1976).

[24]Bone and Ranney (1976, p. 35, Figure 3).

[25]Pomper et al. (1977, p. 72).

about 60 percent in recent elections--do not vote,[26] though
typically there are more candidates from whom to choose than
in the general election.

Spatially, I shall assume that the alienation of a voter
is a direct function of his distance from the position of the
candidate closest to his position. If this distance is sufficiently great, then the voter's alienation overcomes his
desire to vote for the candidate closest to him and he becomes
a nonvoter. In the economist's language, if the demand for a
product (candidate) is elastic (i.e., depends on its price),
that product (candidate) will not be purchased if the price
for a customer (voter) becomes too high (voter is too far
from a candidate's position).

The alienation of voters "too far" from any candidate's
position may contravene findings from our earlier analysis.
For example, alienation will tend to undermine the desirability
of the median/mean in Figure 3, and enhance the desirability of
the two modes in this figure, as the optimal positions in a
two-candidate race.

The reason is that the number of voters alienated a
given distance from the median/mean may be more than the numbers alienated the same distance from either mode. The
decrease in the number of alienated voters at the modes implies
an increase in voter support, making the modal positions more
attractive to the candidates.

Exercise 14. For the bimodal voter distribution given in Exercise 3,
assume that voters will not vote for a candidate if his position on the
0-1 scale is more than 0.1 units from theirs. If a candidate has no opponent, what position will maximize his vote total?

---

[26]Ranney (1972, p. 24, Table 1). On factors that affect turnout in
primaries, see Morris and Davis (1979) and Ranney (1977).

Thus, a bimodal distribution in which alienation is a
factor may induce rational candidates to adopt polarized
positions on the left and right of an issue rather than locate
themselves near the median. While advocates of "responsible"
parties (and candidates) that present clear and distinct
choices to the voters will view this polarization as salutary,
advocates of compromise will not be enamored of the black and
white choices that such polarization entails.

One way that a candidate can reduce his distance from
voters, and possibly avoid the vote-draining effects of
alienation, is to fuzz his position. Given that voters perceive a candidate's ambiguity as favorable to them, a strategy
of ambiguity will increase the broadness of his appeal.

To illustrate the possible advantages of ambiguity,
assume that a candidate's true position is at the center of
the band in Figure 10. If the candidate does not fuzz his
position, assume that the "reach" of this position along the
continuum is that shown as "true" in Figure 10.

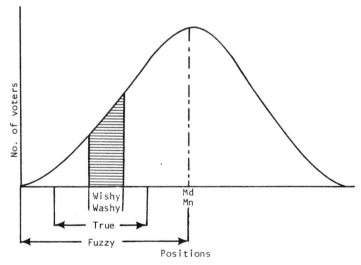

Figure 10. Fuzzy position of a candidate.

If the candidate fuzzes his position, however, he might be able to extend its reach from the left extreme to the median, assuming that voters on the left extreme interpret his position to be the left boundary of the band and voters at the median interpret his position to be the right boundary of the band. On the other hand, if voters, assuming the worst, make the opposite interpretation--the boundaries of the band farthest from them are the actual positions of the candidate--an ambiguous candidate may perversely succeed in contracting (rather than expanding) his support when he fuzzes his true position. Call this interpretation of a candidate's position by voters "wishy-washy" and assume its reach to be only the bandwidth itself, versus the "fuzzy" range, in Figure 10.

Thus, a danger may attend a strategy of ambiguity, depending on what voters perceive to be the actual position of a candidate. Or, given that they recognize the ambiguous strategy of a candidate to be a band rather than a point on the continuum, their choice may then depend on whether they view this ambiguity to represent a desirable flexibility or an undesirable pusillanimity.

Exercise 15. For the unimodal voter distribution given in Exercise 9, assume a candidate's true position is at 0.3. If perceived as "wishy-washy," assume the candidate gains the votes of voters only at 0.3; if "true," he extends his appeal to voters up to a distance of 0.1 units away; if "fuzzy," he extends his appeal still farther to a distance of 0.2 units away. Given the candidate has no opponent, how many votes do these different perceptions by voters yield him?

Apparently, voters have responded to ambiguity differently in different elections. Nobody ever accused Richard Nixon of forthrightness in his 1968 presidential campaign when he said, "I have a plan" to end the war in Vietnam. But, judging from the results of the Republican primaries and the general

election in 1968, more voters believed in his competence to
deal with the Vietnam situation than believed in the more
specific proposals of his opponents.

In contrast, as George McGovern became increasingly
vague about specific proposals he had made in the early Democratic primaries in 1972, and then withdrew his initial
"1,000 percent" support of his vice-presidential choice,
Thomas Eagleton, after the convention, voters began to see
him as irresolute. At the polls, they overwhelmingly chose
the by then better-known quantity, incumbent Nixon, in the
1972 election. Of course, only a few months after this
election, the unravelling yarn of Watergate turned Nixon's
presidential image into a shambles.

Jimmy Carter's positions before and after the 1976 election present an interesting blend in contrasts. During the
campaign he was quite unspecific on a number of issues, but
after his election he developed a number of detailed programs
(e.g., on energy and welfare) that he presented to Congress.
Should he run for reelection in 1980, his campaign strategy
as an incumbent president will undoubtedly less emphasize
moral and spiritual themes and more stress his specific
accomplishments as president.

These examples would seem to indicate that a strategy
of ambiguity may be productive or unproductive, depending on
how the candidate is viewed by the voters. From a spatial
perspective, an ambiguous strategy would seem least risky
for a candidate who tries to push his support toward the
extremes, given that he can also hold onto more moderate
voters with another position near the center. On the other
hand, a candidate squarely but ambiguously in the center is
more likely to have to counter attacks from both his left
and right, which may dissolve his centrist support on both
sides, especially if his opponents can represent his position to be at the boundary of the band farthest from them.

Admittedly, these conclusions are rather speculative, principally because very little is known about what kinds of factors engender support for, or opposition to, fuzzy positions. In the absence of such knowledge, I can make only tentative assumptions about the relationship between ambiguous strategies and voting behavior and indicate the consequences each implies.

I suggested earlier that voter alienation is pervasive, but its implications are not entirely clear, especially in primaries. To begin with, citizens may fail to vote in the early primaries not so much because they find the candidates unattractive as they know very little about them. This might be called indifference due to ignorance: voters may not even know how to bracket the candidates, much less their specific positions.[27] However, as the field narrows in later primaries, and more information is generated about the races in both parties, the positions of candidates--specific or ambiguous--become clarified. Then alienation due to incompatability, which I stressed earlier, may begin more and more to manifest itself.[28]

As early contenders are eliminated and the appeal of the surviving candidates broadens, each will feel less of a need to draw a fine line between himself and the other survivors, who will generally be spaced farther apart along the continuum. Hence, there will be an incentive for a candidate to extend his position from a point to a band to take in voters who otherwise would be alienated because they fall between, or--if situated at the extremes--too far away from, positions that have been eliminated.

---

[27]Three out of five supporters of Eugene McCarthy, the antiwar candidate in the Democratic primary in New Hampshire in 1968, believed that the Johnson administration was wrong on Vietnam because it was too dovish rather than too hawkish--a complete inversion of McCarthy's views. Scammon and Wattenberg (1970, p. 91).

[28]Riker and Ordeshook (1973, pp. 323-330) draw a similar distinction between "indifference" and "alienation," though they use the former concept to refer to a "cross-pressured" voter, not one who simply lacks information.

But then the danger of being seen as wishy-washy or evasive, especially when sharpened by attacks from the opposition, may inspire contraction as well. The frequently observed consequence of buffeting by these contradictory forces is to-and-fro movements as candidates hew to basic positions but at the same time scamper for pockets of support somewhat removed from these positions. It is fascinating to watch this dance performed along the continuum, even if it does not always seem well rehearsed.

9. Political Parties:
Three-Headed Monsters

So far I have used a simple spatial model, which assumes
only a distribution of voters along a left-right continuum,
to analyze the competition of candidates in presidential primaries. In the general election, however, parties become
signficant forces. Hence, it is appropriate now to introduce
possible divergent interests within parties that will compli-
cate the previous analysis. The question to be answered is
what coalition of party interests will form to meet competition from the outside.

American political parties have a colorful history, and
literally millions of words have been written about them and
the candidates who have represented them. Still, their images,
and the way they function in the American political system,
remain somewhat of a mystery, although there is general agreement that the major parties embrace a curious cast of
characters.

In the coalition model to be developed in subsequent
sections, I assume that parties contain three distinguishable

sets of players: (i) professionals, (ii) activists, and
(iii) voters. The professionals are elected officials and
party employees who have an obvious material stake in the
party's survival and well-being. The activists are amateurs--
either voters or candidates--who volunteer their services or
contribute other resources to the party, especially during
elections.[29] The voters, who make up the great mass of the
party, generally do not participate in party activities,
except to vote or possibly make minimal contributions.

It is this mixture of players, each with their own
diverse interests, that makes a party a "three-headed monster"--
not so much because parties are terrifying creatures but rather
because they are so hard to control. That is why it is useful
to think of parties as coalitions of players whose members
somehow must reach agreement among themselves if they are to
be effective political forces.

What complicates the process of reaching agreement is
that the activists tend to take more ideologically extreme
positions than the professionals and ordinary voters. There
are exceptions, of course, but I assume in the subsequent
analysis that activists give their support because they be-
lieve in, or can gain from, the adoption of certain extremist
policies.

Not only do these policies generally give them certain
psychic or material rewards, but they also usually exclude
others from similar benefits. Activists tend to be purists,
and they are not generally satisfied by "something-for-
everything" compromise solutions.

Professionals, on the other hand, are interested in the
survival and well-being of their party, and they do not want
to see its chances or their own future employment prospects
jeopardized by the passions of the activists. Their positions

---

[29]Robertson (1976, pp. 31-33) also introduces activists in his model
of party competition.

generally correspond to those of the median voter, whom they do not want to alienate by acceding to the wishes of the activists.

Yet, by virtue of the large contributions the activists make to the party, activist interests cannot be ignored. The election outcome, I assume, would be imperiled if the professionals, who are mainly interested in winning, lost either the support of the activists or the support of the voters.

What is the outcome of such a medley of conflicting forces? Before possible outcomes can be analyzed, the goals of candidates--what they seek to optimize, given the conflicting interests of the various groups whose support they seek--must be specified.

10. Reconciling the Conflicting Interests

In previous sections I analyzed the positions of candidates in primaries that were both optimal and in equilibrium vis-à-vis one or more other primary candidates. After the nomination of one candidate by each of the major parties at its national convention, the presidential-election game is usually reduced to a contest between only two serious contenders in the general election.

To generate financial support (primarily from activists) and electoral support (primarily from voters) in the general election, I assume that a candidate tries to stake out positions--within certain limits--that satisfy, or at least appease, both activists and voters. To model his decisions in the general election, I shall ignore for now the positions that the other major-party candidate may take. While the positions of a candidate's opponent will obviously determine in part his own positions as the campaign progresses, I assume in the subsequent analysis that a party nominee's top-priority goal after the convention is to consolidate his support within the ranks of his own party.

To satisfy this goal, I assume that a candidate cannot afford to ignore the concerns of either the activists or the voters. Without the support of the former, a candidate would lack the resources to run an effective campaign; without the support of the latter, his appeal would be severely attenuated even if his resources were not.

Consequently, I assume that a presidential candidate seeks to maximize both his resources and his appeal, the former by taking positions that increase his attractiveness to activists and the latter by taking positions that increase his probability of winning among voters.[30] Specifically, if resources (contributed by activists) are measured by the utility (U) activists derive from his positions, and appeal (to voters) by the probability P that these positions--<u>given</u> sufficient resources to make them known--will win him the election, then the goal of a candidate is to take positions that maximize his expected utility (EU), or the product of U and P:

$$EU = U(to\ activists)P(of\ winning\ among\ voters).$$

If effect, the EU calculation provides a measure of the combined activist and voter support that candidates can generate from taking particular positions in the general election.

Maximization of EU implies seeking a compromise satisfactory to both the activists and the voters. Normally, this compromise will be aided by professionals who seek to reconcile the conflicting interests of the two groups. In section 11, I shall show what form this reconciliation may take, depending on the nature of the conflicting interests that divide the activists and the voters.

---

[30]For other perspectives on goals, see Schlesinger (1975) and Wittman (1973). On difficulties parties now face, see Pomper (1977, pp. 13-38) and Ranney (1975).

11. Optimal Positions
in a Campaign

For simplicity, assume that the campaign involves a single issue, and the positions on this issue that a candidate of the left-oriented party may take range from the left extreme (LE) to the median (Md), as shown in Figure 11. Assume further that the utility (measured along the vertical axis) that activists derive from the positions a candidate takes along the horizontal axis falls linearly from a high of 1 at LE to a low of 0 at Md. On the other hand, assume that the probability of winning (also measured along the vertical axis) varies in just the opposite fashion, starting from a low of 0 at LE and rising to a high of 1 at Md.[31]

---

[31] If a candidate's opponent also adopts a position at Md, then the candidate's P at Md will be 0.5 instead of 1.0, assuming the activist support (resources) of both candidates are the same at Md. Although the actual value of a candidate's P at Md--dependent on his opponent's behavior--does not affect the maximization of EU, it may affect strategy choices in a manner to be discussed later.

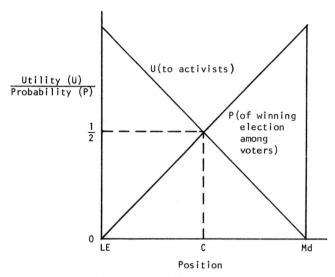

Figure 11.   Utility and probability of
candidate positions.

I assume that the maximum probability of winning cannot
be attained, however, unless adequate resources are contributed by activists to publicize the nominee's positions.
Since a left-oriented activist derives 0 utility from a
candidate who takes the median position, it seems reasonable
to assume that no resources will be contributed to a left-
oriented candidate whose position is at Md.

A candidate increases his resources, but decreases his
probability of winning, as he moves toward the left extreme.
Clearly, if he moves all the way left to LE, P = 0, just as
U = 0 at Md. Thus, a candidate who desires to maximize EU
would never choose positions at LE or Md where EU = 0.

In fact, it is possible to show that the optimal position of a candidate is at the center (C) of Figure 11, i.e.,
the point on the horizontal axis midway between LE and Md
where the lines representing U and P intersect. Since this

point is also midway between 0 and 1 on the vertical axis--
at the point 1/2--

$$EU = (\tfrac{1}{2})(\tfrac{1}{2}) = \tfrac{1}{4} .$$

There is no other point on the horizontal axis at which
a candidate can derive greater EU. Consider, for example,
the point midway between C and Md, where U = 1/4 and P = 3/4.
At this position,

$$EU = (\tfrac{1}{4})(\tfrac{3}{4}) = \tfrac{3}{16},$$

which is less than EU = 1/4 at C.

The optimality of position C in Figure 11 may be upset
if U and P are not linear functions of a candidate's position
(i.e., functions that can be represented by straight lines)
but instead are curves like those shown in Figure 12. As in
Figure 11, the utility of a candidate's position decreases,

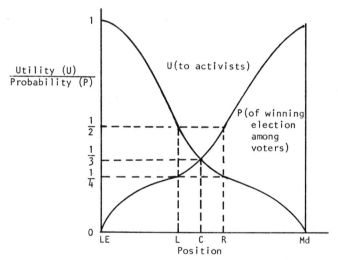

Figure 12. Nonlinear utility and probability
functions.

and the probability of his position being winning increases, as the candidate moves from LE to Md. Now, however, since U and P are not linear functions of a candidate's position along the horizontal axis, the point of intersection of the P and U curves at C on the horizontal axis may no longer be optimal.

To illustrate this proposition, calculate EU at C and at points to the left and right of C. Clearly, at C in Figure 12,

$$EU = (\tfrac{1}{3})(\tfrac{1}{3}) = \tfrac{1}{9} = 0.111,$$

but at L (to the left of C)

$$EU = (\tfrac{1}{2})(\tfrac{1}{4}) = \tfrac{1}{8} = 0.125,$$

and at R (to the right of C),

$$EU = (\tfrac{1}{4})(\tfrac{1}{2}) = \tfrac{1}{8} = 0.125.$$

Hence, given the nonlinear utility and probability functions shown in Figure 12, a candidate can do better by taking a position either to the left or to the right of C.

The exact positions along the horizontal axis which maximize EU for a candidate will depend on the shape of the U and P curves. These optimal positions can be determined from the equations that define the curves, but since there is no empirical basis for postulating particular functional relationships between candidate positions and U and P, I shall not pursue this matter further here.

The main qualitative conclusion derived from this analysis is that there is nothing sacrosanct about the center position C. Depending on the shape of the U and P functions, a candidate may do better--with respect to maximizing EU--by moving toward LE, toward Md, or in either direction.

Exercise 16. Try drawing different-shaped U and P curves to illustrate different conclusions about the location of optimal positions.

Whatever the shape of the U and P functions, however, if P = 0 at LE and U = 0 at Md, the positions at LE and Md will never be optimal since EU = 0 in either case. But as long as U decreases monotonically from LE to Md (i.e., does not change direction by first decreasing and then increasing), and P increases monotonically from LE to Md, any points in between LE and Md may be optimal, depending on the shape of the U and P curves.

Exercise 17. What can one say if the curves are not monotonic? Is there any reason for assuming that these curves may not be monotonic?

If these curves are symmetric (i.e., mirror images of each other, as in Figure 12), there may be two optimal positions, one on each side of C. Yet symmetry is not a sufficient condition for there to be more than one optimal position: the straight lines in Figure 11 are symmetric, but the only position along the horizontal axis where EU is maximized is at C.

What are the implications of this analysis? If activists prize "extremeness," and ordinary voters prize "moderation," then any position in between may be optimal for a candidate who desires to maximize some combination of his resources (from activists) and his electoral support (from voters). More surprising, there may be different optimal positions, one more favorable to the activists and one more favorable to the voters, as illustrated in Figure 12.

12. Empirical Examples of Different Optimal Positions in Campaigns

So far I have shown how a model might offer an explanation--in terms of a candidate's desire to maximize EU-- for the optimality of different positions in a campaign. The utility and probability functions that I postulated may, depending on their shape, push candidates toward an extreme position (left extreme in my example), the median position, or a center position somewhere in between.

In recent presidential campaigns, it is possible to observe a variety of positions that nominees of both major parties have adopted. Barry Goldwater, the 1964 Republican nominee, and George McGovern, the 1972 Democratic nominee, provide the best examples of candidates who took relatively extreme positions in their campaigns. Both candidates had strong activist support from the extremes of their parties in the primaries, which they almost surely would have lost had they tried to move too far toward the median voter in the general election. In addition, given the moderate

opposition both candidates faced from relatively strong
incumbents in the general election, neither Goldwater nor
McGovern probably stood much chance of picking up many voters
near the median had he tried to shift his early extremist
positions very much.

If Goldwater had run against John Kennedy rather than
Lyndon Johnson in 1964, however, he probably would have been
a viable candidate. He could have carried all the South and
West and some of the Midwest and, conceivably, might have
won. Against Johnson, though, he was a loser because he and
Johnson appealed in great part to the same interests, while
the old Kennedy voters were stuck with Johnson. Goldwater
planned his strategy with Kennedy alive and could not jetti-
son it after Kennedy was assassinated.

By comparison, McGovern's early extremist positions
were no match from the beginning against Nixon's middle-of-
the road positions. When, in desperation, McGovern attempted
to moderate some of his early positions, he was accused of
being "wishy-washy" and probably suffered a net loss in
electoral and financial support (see section 8).

In general, if the utility for activists falls off
rapidly, and the probability of winning increases only slowly,
as a candidate moves toward the median, his optimal position
will be near the extreme. Such a position gains more in
resources than he loses in probability of winning compared
with a position near the median. With this trade-off in mind,
both Goldwater and McGovern seem to have acted rationally with
respect to the maximization of EU, though McGovern seems to
have been more willing to sacrifice activist support to increase his chances of winning.

The incumbent presidents that Goldwater and McGovern
faced, Lyndon Johnson and Richard Nixon, had more moderate
activist supporters who were less disaffected by "middle-of-
the-road" politics. Not only could these incumbents afford
to move toward the median voter and still count on significant activist support, but, because of the extreme positions

of their opponents, they could probably rapidly increase the number of their moderate supporters with such a strategy.

However, as James S. Coleman has pointed out, if an incumbent already has greater a priori strength than his opponent--and his opponent magnifies the discrepancy in strength by adopting an extremist position--the incumbent will not significantly improve his (already high) probability of winning by moving farther away from the other extremist position and toward the median.[32] Against such an opponent, therefore, an incumbent with a large built-in advantage from the start has little incentive to move toward him. Thus, extremist positions, especially when there is an a priori difference in electoral strength (e.g., when a nonincumbent runs against an incumbent), will tend to reinforce each other: both candidates will be motivated to adopt relatively extreme positions, because movement by one candidate toward the other more decreases his activist support than it increases his probability of winning.

The problem with this conclusion is that it seems to have little empirical support. The Goldwater-Johnson and McGovern-Nixon races did not produce extremists on both sides but only on one. In fact, if one candidate's position diverges sharply from the median, as did those of Goldwater and McGovern, there seems a tendency for his opponent to move toward his position rather than in the opposite direction.

This behavior is explained quite well by our earlier spatial models (see, in particular, section 4), but it is difficult to derive it from the goal of maximization of EU in which P is one factor. After all, if P is already high for a strong incumbent running against an opponent who adopts an extremist position (for reasons given earlier), why should the incumbent move toward his opponent if this movement has little effect on P and may lower U at the same time?

---

[32]Coleman (1973).

The answer seems to lie in the fact that some candidates seem to be as interested in the absolute size of their majorities as in winning. That is, they desire large majorities at least as much as victory itself. If this is the case, then movement toward an extremist opponent can be explained by the fact that this movement steadily increases a strong candidate's vote total even if it does not significantly alter his probability of winning.

Both Johnson and Nixon ran campaigns which strongly indicate that, even with victory virtually assured months before the election, they wanted more than victory: they desired to pile up huge majorities by whatever means they had at their disposal (including misrepresentation of their positions and those of their opponents). Although both incumbents succeeded in crushing their opponents in their respective elections, both were later driven from office by a welter of forces that I have analyzed elsewhere.[33]

If the goals presidential candidates seek to maximize preclude both candidates from diverging from the median--and may encourage convergence, as in the 1960 and 1968 presidential elections--then it is unlikely that one of the major parties can be written off the national political scene for very long. Indeed, in recent presidential elections, there has been a steady alternation of ins and outs: no party since World War II has held office for more than two consecutive terms. This alternation of ins and outs was not nearly so steady before the post-war era, with one or the other party on occasion holding sway for a generation or more.

---

[33]Brams (1975, chap. 6) and Brams (1978, chap. 4).

13. Multiple Issues in a Campaign

The dance along the continuum alluded to at the end of section 8 may be complicated if there is more than one issue, or policy dimension, on which candidates take positions and voters base choices. For then a voter's distance from a candidate's position must be measured in two- or higher-dimensional space, and optimal positions of candidates with respect to different distributions of voter attitudes become considerably harder to determine.[34]

The problem is rendered more difficult if voters weight the various issues differently. Some voters, for example, may attribute more importance to a candidate's position on economic issues than foreign policy issues, while others may reverse this attribution. In general, the salience of issues for voters, or the relative importance they attach to

---

[34]A geometric treatment of optimal positions in two dimensions is given in Tullock (1967, chap. 4).

candidate positions on them, obviates any simple extension of the one-dimensional spatial analysis to higher dimensions, especially when salience is correlated with the attitudes of voters on issues.[35] In addition, the interrelatedness of some issues may invalidate their representation as independent dimensions on which candidates are separately evaluated.

Despite these difficulties, it is important to try to analyze some elementary consequences of multi-issue campaigns. For this purpose, consider a simple example of a campaign in which there are just two issues, X and Y.

Assume that each candidate can take only one of two positions on each issue (e.g., for or against), which I designate as x and x', y and y'. Altogether, there are four possible platforms, or sets of positions on both issues, that a candidate can adopt: xy, x'y, y'x, or x'y'.

Assume that the electorate consists of three voters, and their preferences for each of the platforms are as shown in Table 1.[36] For each voter, the first platform in parentheses is his most preferred, the second his next-most preferred, and so on.

TABLE 1

PREFERENCES OF THREE VOTERS FOR PLATFORMS

Voter	Preference
1	(xy, xy', x'y, x'y')
2	(xy', x'y', xy, x'y)
3	(x'y, x'y', xy, xy')

---

[35] Jackson (1973).

[36] This example is taken from Hillinger (1971); see also Kadane (1972) for an analysis of the effects of combining different alternatives.

Assume that there are just two candidates, and one is elected if a majority of voters (two out of three) prefers his platform to that of the other candidate. What platform should a candidate adopt if his goal is to get elected?

To answer this question, one might start by determining which position on each issue would be preferred by a majority if votes were taken on the issues separately. Since x is preferred to x' by voters 1 and 2, and y is preferred to y' by voters 1 and 3 (compare the first preferences of the voters in Table 1), it would appear that platform xy represents the strongest set of positions for a candidate.

But this conclusion is erroneous in the example here. Despite the fact that a majority prefers positions x and y were the issues voted on separately, platform x'y' defeats platform xy since it is preferred by a majority (voters 2 and 3). Thus, a platform whose positions, when considered separately, are both favored by a majority may be defeated by a platform containing positions that only minorities favor. A recognition that a majority platform may be constituted from minority positions is what Downs argued may make it rational for candidates to construct platforms that appeal to "coalitions of minorities."[37]

The divergence between less-preferred individual positions and a more-preferred platform that combines them depends on the existence of a paradox of voting.[38] In this example, this means that there is no platform that can defeat all others in a series of pairwise contests. As shown by the arrows in Figure 13, which indicate majority preferences

---

[37] Downs (1957, chap. 4).

[38] Hillinger (1971, p. 565) claims this is not the case, but this is refuted in Miller (1975, p. 110). A paradox of voting also underlies what has been called the "Ostrogorski paradox," which is essentially the same as that illustrated in the text. See Rae and Daudt (1976). For a description of, and review of the literature on, the paradox of voting, see Brams (1976, chap. 2).

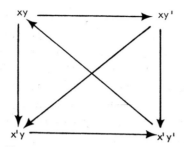

Figure 13. Cyclical majorities
for platform voting.

between pairs of platforms, every platform that receives
majority support in one contest can be defeated by another
majority in another contest. For this reason, the majorities that prefer each platform are referred to as <u>cyclical</u>
<u>majorities</u>.

Exercise 18. If xy and x'y' were interchanged in the preference ranking
of voter 1, would majorities by cyclical? If not, which platform would
defeat all others in a series of pairwise contests?

Exercise 19. If xy and x'y' were interchanged in the preference ranking
of voter 2, would majorities be cyclical? If not, which platform would
defeat all others in a series of pairwise contests?

Exercise 20. Now assume that xy and x'y' are interchanged in the
preference rankings of both voter 1 and voter 2. Would majorities be
cyclical? What if these platforms were interchanged in the preference
rankings of all three voters?

Exercise 21. Prove that if majorities are cyclical, they will remain
cyclical if two platforms are interchanged in the preference rankings of
all voters.

The main conclusion derived from the simple example in
this section is that there may be no set of positions that a
candidate can adopt on two (or more) issues that is invulnerable: any set of positions that one candidate takes can

be defeated by a different set adopted by another candidate. This means that, without any shift in the preferences of voters, a candidate running on a given platform could win an election in one year, and lose it in the next, depending on the positions his opponent took.[39] This fact helps to explain the importance that candidates attach to anticipating an opponent's positions so that they can respond with a set that is more appealing to the voters.

Of course, some candidates try to avoid this problem by being intentionally vague about their positions in the first place, as Downs pointed out.[40] But this strategy of ambiguity may lead to its own problems, as I showed in section 8.

By now it should be evident why primaries, and the general election later, so often seem to yield topsy-turvy outcomes in presidential races. The strongest theoretical result discussed in this monograph--the stability and optimality of the median in a two-candidate election--can be undermined if there is more than one issue on which candidates take positions. Indeed, no set of positions will be stable if there exists a paradox of voting, nor will any set be optimal in the sense of guaranteeing a particular outcome whatever the positions of one's opponent. In fact, contrary to expectations, one's best set of positions on issues in a race may be the minority positions on the issues considered separately, depending on the positions of one's opponent.

These findings do not depend on the exact nature of the underlying distributions of attitudes of voters or the precise location of candidates with respect to his distribution. They depend only on qualitative distinctions (dichotomous positions of candidates, ordinal preferences of voters) and are, therefore, of rather general theoretical significance whatever the quantitative characteristics of a race are.

---

[39]Frohlich and Oppenheimer (1978, p. 135).
[40]Downs (1957, chaps. 8 and 9).

Probably the best advice to take from the analysis in this section is negative: avoid reading too much into spatial analysis based on a single issue if there may be other issues of significance in a campaign. Multiple issues greatly complicate--and may ultimately confound--single-issue spatial analysis, as the paradoxical findings in this section illustrate. Nevertheless, it is important to try to link candidate positions and voter attitudes, and spatial analysis provides a useful framework within which to relate these characteristics in both a series of primaries and a single election.

14. Summary and Conclusion

In this monograph, some of the hurdles that presidential
candidates face first in state primaries and then in the
general election were explored. In the analysis of primaries,
I assumed that the principal goal of a candidate is to avoid
elimination, if not win; by contrast, voters want to maximize
their satisfaction on the issue they consider most important
by choosing the candidate whose position is closest to theirs.
The spatial games candidates play to try to maximize their
appeal to voters were the focus of most of the analysis of
primaries.

I first considered the case of two candidates who vie
for the most favorable position along a left-right continuum
in a single-issue campaign. I showed that, whatever the
distribution of voter attitudes on the issue, the median is
best for two reasons: (i) it is optimal--there is no other
position that guarantees a candidate a better outcome; (ii) it
is in equilibrium--once chosen by both candidates, neither
would have an incentive to depart unilaterally from it. A

corollary of this finding is that an "average" position (at the mean) is not optimal or in equilibrium if the distribution of voter attitudes is skewed to the left or right and the median, as a consequence, does not coincide with the mean.

In multi-candidate races, not only does the median lose its appeal but any positions (not necessarily the same) that two candidates might take are vulnerable to the entry of additional candidates. I suggested that this fact helps to explain why so many candidates are motivated to enter the early primaries and try to displace other nearby candidates on the left-right continuum.

The initial competition among liberal, moderate, and conservative candidates in each segment of the distribution leads to a winnowing out of marginal candidates. This elimination process tends to favor more moderate-leaning candidates on the left and right, who then can effectively challenge a centrist candidate in the middle. The results of recent primary campaigns suggest that a liberal or conservative candidate who receives the support of as few as 25 percent of party voters in all primaries can squeeze out one or more centrist candidates in the final competition.

A well-known candidate who can afford to await the results of the early primaries before making his announcement of candidacy can benefit from knowing the early survivors' positions. I illustrated the advantages of a delayed announcement, and the importance of timing in a campaign generally, by showing how a moderate could maximize his support by moving toward the less extreme of his opponents on the left or right after they had committed themselves, given a symmetric, unimodal distribution of voter attitudes.

I next showed that a candidate who fuzzes his position might be either helped or hurt, depending on the voters' perception of his true position. Or, if voters correctly perceive his position to be one of ambiguity, their evaluation of his competence to deal later with the issue at hand will likely seal his electoral fate.

I suggested that voter alienation may induce candidates to fuzz their positions in order to try to embrace a wider swath of voters. Alienation may also push candidates toward modal positions, where voters are most concentrated, because the voters who are alienated by being too distant from a mode will generally be fewer than those too distant from other points in the distribution. In particular, if the distribution of voter attitudes is bimodal, voter alienation will encourage a polarization of candidate positions on the left and right.

In the general-election coalition model, parties were not assumed to be unitary actors but rather an amalgam of diverse interests. I postulated that presidential candidates would seek to maximize a combination of activist support (resources) and electoral support (probability of winning), which were assumed to move in opposite directions with respect to a candidate's position on an issue. That is, as a candidate moves toward the median position, he alienates his activist supporters but increases his probability of winning; on the other hand, as he moves toward an extreme position, the reverse trade-off occurs.

I showed that a candidate who wishes to maximize his expected utility (i.e., activist utility times probability of winning) should take a position between the median and an extreme position--exactly where depending on the shape of the utility and probability curves. I demonstrated that there may be more than one optimal position for a candidate--one near the median, the other near an extreme--and also showed how optimal positions might change if a candidate's goal included a desire not just to win but also to maximize his vote total. Optimal positions derived from this modified goal seemed to be consistent with the campaign behavior of candidates in recent presidential elections.

Finally, I showed how multiple issues may upset the calculations of one-dimensional spatial analysis. Specifically, the existence of a paradox of voting will make every platform vulnerable to challenges, which means that no positions are in equilibrium, even when there are just two candidates. Also, the fact that there may be no unconditionally best, or optimal, platform means that platforms that comprise minority positions on two or more issues considered separately may defeat platforms comprising majority positions on the separate issues. For these reasons, I concluded that findings derived from one-dimensional spatial models must be treated with caution if there is more than one issue in a campaign on which the positions of candidates determine the behavior of voters.

15. Appendix

In this Appendix I offer a somewhat more formal development of the results discussed informally in section 5 and in the answers to exercises 6 and 8 of this section. Assume the following in a single-issue political race:

1. There is a left-right ideological dimension under-lying this issue along which candidates take positions.
2. Each voter has a most-preferred position on this dimension.
3. Each voter has one vote and always casts it for the candidate whose position is closest to his most-preferred position.
4. The candidate with the most votes wins (plurality voting).

To begin the analysis, assume that there are two candidates, a liberal (L) and a conservative (C), whose positions on the left-right ideological dimension are known. Designate

positions on this dimension by the real variable x, and
assume voters are distributed over the interval $a \leq x \leq b$
according to continuous density function f(x), where
$f(x) > 0$ if $x \neq a$ and $x \neq b$.

Since f(x) is assumed to be a continuous density function,
$\int_a^b f(x)dx = 1$. Although I shall not give a probabilistic
interpretation to f(x), it is convenient to assume this kind
of distribution of voters in order to be able to derive
numerical results that indicate fractions of the electorate
falling between points on the left-right continuum.

Assume x = M is the median of the distribution, x = L is
the position (as well as name) of the liberal candidate, where
$a \leq L < M$, and x = C is the position (as well as name) of the
conservative candidate, where $M < C \leq b$. I shall now prove
that if fewer than 1/3 of the electorate lies between L and C--
between each of whom and M there are the same (nonzero) number
of voters--there is no position that a third candidate can
take along the left-right dimension that is winning.

In other words, a third candidate cannot knock out <u>both</u>
the original entrants and win the election if the original
entrants straddle the median in such a way that $\leq$ 1/6 of the
electorate lies between each and the median. While I assume
that the same number of voters ($\leq$ 1/6) lies between M and L
and between M and C, I assume nothing about the shape of the
voter distribution except that f(x) is always positive in the
domain $a < x < b$. These results are summarized in

THEOREM 1. <u>Let</u> x = L <u>and</u> x = C <u>be the positions of the</u>
<u>liberal and conservative candidates, respectively, and let</u>
x = M <u>be the median of continuous density function</u> $f(x) > 0$
<u>that defines the distribution of voter positions over the</u>
<u>interval</u> $a < x < b$. <u>If</u>

$$0 < \int_L^M f(x)dx = \int_M^C f(x)dx \leq 1/6,$$

there is no position x = X that some third candidate X can
take that is winning.

Proof. For X to be winning, he must receive more votes
than both L and C. There are four possible sets of positions
he can take along the left-right continuum: (1) between a and
L; (2) between L and C; (3) between C and b; (4) at L or C.
Consider each in turn:

1. a ≤ X < L: Clearly, X maximizes his vote total by
   taking a position just to the left of L; any other
   position, closer to a, would mean that he would lose
   votes to L since some voters falling between them
   would be closer to L. But his vote total will always
   be less than C's because C will gain not only all the
   votes to his right (the same number as to the left of
   L that X receives) but also some votes between L and
   C that X will not receive because L is just to his
   right.

2. L < X < C: Since the number of votes between L and
   C is ≤ 1/3, L and C would receive > 1/3 of the votes
   and thereby both surpass the vote total of X.

3. L < X ≤ b: Reasoning analogous to (1) above, but
   with left and right reversed.

4. X = L or X = C: The candidate whose position X
   does not take would have > 1/3 of the vote, whereas
   X and the candidate whose position he takes would
   split the remainder of the vote, each obtaining < 1/3.

Hence, there is no position x = X that will ensure X more votes
than one or both the original entrants. Q.E.D.

Note that X can always displace either L or C by taking
a position just to his left or right, respectively. But in
so doing, he always ensures the other original candidate some
portion of the votes in the middle between L and C--in addition
to those to his left or right--that makes the other candidate
victorious.

Theorem I demonstrates that in a noncooperative three-person, zero-sum game, a rational player may do worse by choosing a strategy after the other players, which is never true in two-person, zero-sum games. In the particular spatial game I have described, the player choosing a position last will always lose, vis-à-vis at least one other player, if the conditions of the theorem are met.

It is easy to show that a relaxation of any of the conditions of the theorem could lead to a win for X. In particular:

1. If $\int_{L}^{M} f(x)dx = \int_{M}^{L} f(x)dx \not> 0$, i.e., if L = M = C, then X could take a position just to the left or right of the median and capture (essentially) 1/2 of the vote, with L and M splitting the remaining 1/2, or receiving 1/4 each.

2. If $\int_{L}^{M} f(x)dx = \int_{M}^{C} f(x)dx > 1/6$, and the > 1/3 votes in the center between L and C were highly concentrated around a mode, X could capture (essentially) all of them by taking a position at the mode, with L and C receiving < 1/3 each.

3. If $\int_{L}^{M} f(x)dx \neq \int_{M}^{C} f(x)dx$, either the number of voters between a and L would be greater than the number between C and b, or vice versa. Without loss of generality, assume the former is the case. Then by taking a position just to the left of L, X would receive > 1/3, and L < 1/3, of the vote. But X could also receive more votes than C, and hence win, if C captured too few votes in the center (e.g.,

because almost all voters in the center were closer to L than C) to augment the < 1/3 to his right.

Thus two candidates, equal numbers of voters distant from the median, cannot both be knocked out by a third candidate as long as they are separated by fewer than 1/3 of the electorate. The "1/3 separation obstacle," however, is no barrier to the displacement of both L and C should a fourth candidate Y also enter the race.

THEOREM 2. Against two candidates L and C, there are always positions third and fourth candidates X and Y can take that ensure that either X or Y wins, unless L and C take positions such that the numbers of votes L or C gains to his left and right are exactly equal. In this case, X or Y can still at least tie L or C for the win.

Proof. Consider the positions of X and Y that are alongside L and C, respectively. Either X can gain more votes by being just to the left of L or just to his right, and similarly for Y with respect to C, unless the numbers of votes L or C gains to his left and right are exactly equal. Assume that X and Y choose such "straddling" positions to maximize their vote totals. (Since these straddling positions are essentially the positions of L and C, already known, maximization by X and Y is independent of the position the other new entrant takes, given that it is a straddling position.) Because these maximizing straddling positions result in X and Y's each receiving more votes than L and C, respectively, L and C will each be displaced by one of the two new entrants, one of whom necessarily wins.

To show what might happen in the exceptional case stated in the theorem, suppose, for example, $f(x) = 1$, $0 \le x \le 1$, and $L = 1/4$ and $C = 3/4$. If X is the third candidate to enter, he can do no better than take the same position as L, thereby splitting 1/2 the total vote with him, or receiving 1/4 of the total. Y will then be indifferent between taking any position x, $1/4 < x \le 3/4$, which will give him 1/4 of the total. However,

only at x = 3/4 will he limit C to 1/4 of the total (and allow
X and L 1/4 each, too), thereby creating a four-way tie.
(Similarly, if X had not earlier taken a position at L, then
L or C would win no matter what Y did.) Thus, if L or C gains
the same numbers of votes to his left and right, X or Y can
still guarantee a tie by taking a position exactly at L or C.
Q.E.D.

If the numbers of votes L or C gains to his left and right
are exactly equal, X or Y may, of course, do worse--lose to L
or C--if either does not occupy the same positions as L and C
do. On the other hand, it is also possible to find examples
in which X or Y can win when L or C gains equal numbers of
votes to his left and right, but these in general will require
coordination between X and Y in a cooperative game. Since the
"equal numbers" condition is a stringent one and, moreover,
does not always render L or C unassailable, it is reasonable to
expect that L or C will, for all practical purposes, be vulnerable to challenges from two new candidates, X and Y.

In summary, I have shown that if two candidates positions
on each side of the median are separated from it by equal
numbers of voters who together constitute 1/3 of the electorate,
the candidates can collectively withstand the challenge of a
third candidate but not the simultaneous challenge of a third
and fourth candidate. These results are independent of the
distribution of the voters on a left-right ideological dimension.

16. References

Asher, Herbert B. (1976). Presidential Elections and American
Politics: Voters, Candidates, and Campaigns since 1952.
Homewood, Illinois: Dorsey Press.

Bone, Hugh A., and Austin Ranney (1976). Politics and Voters.
New York: McGraw-Hill Book Co.

Brams, Steven J., and Philip D. Straffin Jr. (1979). "The
Entry Problem in a Political Race." New York University.

Brams, Steven J. (1975). Game Theory and Politics. New
York: Free Press.

Brams, Steven J. (1976). Paradoxes in Politics: An Introduction to the Nonobvious in Political Science. New
York: Free Press.

Brams, Steven J. (1978). The Presidential Election Game.
New Haven: Yale University Press.

Coleman, James S. (1973). "Communications." American Political Science Review 67, no. 2 (June): 567-569.

Downs, Anthony (1957). An Economic Theory of Democracy. New York: Harper and Row.

Flanigan, William H., and Nancy H. Zingale (1975). Political Behavior of the American Electorate, 3d ed. Boston: Allyn and Bacon.

Frohlich, Norman, and Joe A. Oppenheimer (1978). Modern Political Economy. Englewood Cliffs, New Jersey: Prentice-Hall.

Frohlich, Norman, Joe A. Oppenheimer, Jeffrey Smith, and Oran R. Young (1978). "A Test of the Downsian Voter Rationality: 1964 Presidential Voting." American Political Science Review 72, no. 1 (March): 178-197.

Hillinger, Claude (1971). "Voting on Issues and on Platforms." Behavioral Science 16, no. 6 (November): 564-566.

Hotelling, Harold (1929). "Stability in Competition." Economic Journal 39, no. 153 (March): 41-57.

Jackson, John E. (1973). "Intensities, Preferences, and Electoral Politics." Social Science Research 2, no. 3 (September): 231-246.

Kadane, Joseph B. (1972). "On Division of the Question." Public Choice 13 (Fall): 47-54.

Kelley, Stanley, Jr., and Thad Mirer (1974). "The Simple Act of Voting." American Political Science Review 67, no. 2 (June): 572-591.

Key, V.O., Jr., with the assistance of Milton C. Cummings, Jr. (1966). The Responsible Electorate: Rationality in Presidential Voting, 1936-1960. Cambridge, Massachusetts: Belknap Press.

Lerner, A.P., and H.P. Singer (1937). "Some Notes on Duopoly and Spatial Competition." Journal of Political Economy 45, no. 2 (April): 145-186.

Levin, Murray (1960). The Alienated Voter: Politics in
Boston. New York: Holt, Rinehart, and Winston.

McKelvey, Richard D. (1975). "Policy Related Voting and
Electoral Equilibria." Econometrica 43, no. 5-6
(September-November): 815-843.

Margolis, Michael (1977). "From Confusion to Confusion--
Issues and the American Voter (1956-1972)." American
Political Science Review 71, no. 1 (March): 31-43.

Miller, Nicholas R. (1975). "Logrolling and the Arrow
Paradox: A Note." Public Choice 21 (Spring): 110.

Miller, Warren E., and Teresa E. Levitin (1976). Leadership
and Change: Presidential Elections from 1952 to 1976.
Cambridge, Massachusetts: Winthrop Publishers.

Morris, William D., and Otto A. Davis (1979). "The Sport
of Kings: Turnout in Presidential Preference Primaries."
American Political Science Review (forthcoming).

Nie, Norman H., Sidney Verba, and John R. Petrocik (1976).
The Changing American Voter. Cambridge, Massachusetts:
Harvard University Press.

Niemi, Richard G., and Herbert F. Weisberg (eds.) (1976).
Controversies in American Voting Behavior. San
Francisco: W.H. Freeman and Co.

Ordeshook, Peter C. (1976). "The Spatial Theory of Elections:
A Review and a Critique," in I. Bridge and I. Crewe (eds.),
Party Identification and Beyond. New York: John Wiley
& Sons, pp.285-313.

"Political Alienation in America" (1976). Society 13, no. 5
(July/August): 18-57.

Pomper, Gerald S. (1977). "The Decline of Partisan Politics,"
in Louis Maisel and Joseph Cooper (eds.), The Impact of
the Electoral Process. Beverly Hills, California: Sage
Publications, pp. 13-38.

Pomper, Gerald (1975). Voters' Choice: Varieties of American Electoral Behavior. New York: Dodd, Mead & Co.

Pomper, Gerald M., Richard W. Boyd, Richard A. Brody, and John H. Kessel (1972). [Articles, comments, and rejoinders.] American Political Science Review 66, no. 2 (June): 415-470.

Pomper, Gerald et al. (1977). The Election of 1976: Reports and Interpretations. New York: David McKay Co.

Rae, Douglas W., and Hans Daudt (1976). "The Ostrogorski Paradox: A Peculiarity of Compound Majority Decision." European Journal of Political Research 4 (September): 391-398.

Ranney, Austin (1975). Curing the Mischiefs of Faction. Berkeley, California: University of California Press.

Ranney, Austin (1977). Participation in American Presidential Nominations, 1976. Washington, D.C.: American Enterprise Institute for Public Policy Research.

Ranney, Austin (1972). "Turnout and Representation in Presidential Primaries." American Political Science Review 66, no. 1 (March): 21-37.

Riker, William H., and Peter C. Ordeshook (1973). An Introduction to Positive Political Theory. Englewood Cliffs, New Jersey: Prentice-Hall.

Robertson, David (1976). A Theory of Party Competition. London: John Wiley & Sons, Ltd.

Scammon, Richard M., and Ben J. Wattenberg (1970). The Real Majority: An Extraordinary Examination of the American Electorate. New York: Coward, McCann and Geoghegan.

Schlesinger, Joseph A. (1975). "The Primary Goals of Political Parties: A Clarification of Positive Theory." American Political Science Review 69, no. 3 (September): 840-849.

Shepsle, Kenneth A. (1974). "Theories of Collective Choice," in Cornelius P. Cotter (ed.), Political Science Annual, V: An International Review. Indianapolis: Bobbs-Merrill Co., pp. 4-77.

Smithies, A. (1941). "Optimum Location in Spatial Competition." Journal of Political Economy 49, no. 3 (June): 423-429.

Stokes, Donald (1963). "Spatial Models of Party Competition." American Political Science Review 57, no. 2 (June): 368-377.

Strong, Donald S. (1977). Issue Voting and Party Realignment. University, Alabama: University of Alabama Press.

Taylor, Michael (1975). "The Theory of Collective Choice," in Fred I. Greenstein and Nelson W. Polsby (eds.), Handbook of Political Science: Micropolitical Theory, vol. 3. Reading, Massachusetts: Addison-Wesley Publishing Co., pp. 413-481.

Tullock, Gordon (1967). Toward a Mathematics of Politics. Ann Arbor, Michigan: University of Michigan Press, 1967.

Wittman, Donald (1977). "Candidates with Policy Preferences: A Dynamic Model." Journal of Economic Theory 14, no. 1 (February): 490-498.

Wittman, Donald (1976). "Equilibrium Strategies by Policy Maximizing Candidates." Mimeographed, University of Chicago.

Wittman, Donald (1973). "Parties as Utility Maximizers." American Political Science Review 66, no. 2 (June): 490-498.

Wright, James D. (1976). The Dissent of the Governed: Alienation and Democracy in America. New York: Academic Press.

17. Answers to
Selected Exercises

1. Assume a candidate's opposition-optimal position is not
   adjacent. Then the portion of voters whose positions
   are between the two candidates will be divided between
   them. But since the first candidate could gain all the
   votes of these in-between voters, and lose none, by moving to a position adjacent to, but not past, his opponent,
   a nonadjacent position can never be opposition-optimal.
   Hence, only an adjacent position can be opposition-optimal.

2. In Exercise 1 I established that a necessary condition
   for a candidate's position to be opposition-optimal was
   that it be adjacent to his opponent's; otherwise, the
   candidate could always gain more votes by moving to an
   adjacent position. Now unless an opponent is at one extreme of the continuum, there are two adjacent positions--
   one to the left and the other to the right of the
   opponent--and one is superior to the other if one's
   opponent is not at the median. Clearly, the adjacent
   position that maximizes a candidate's vote total will be

that which is closer to the median because it includes the votes of all voters on the side of the median his opponent does not occupy (50 percent) plus the votes of all voters on his opponent's side from the median up to his opponent's position.

3. The median position is 0.6 since 8 voters lie to the left and 8 to the right. The mean is

$$Mn = \frac{1}{19}[1(.1) + 3(.2) + 2(.3) + 2(.5) + 3(.6) + 6(.8) + 2(.9)]$$

$$= \frac{1}{19}(10.7) \simeq 0.56.$$

4. The median position is 0.5 since 11 voters lie to the left and 11 to the right. The mean is

$$Mn = \frac{1}{25}[2(0) + 3(.2) + 4(.3) + 3(.4) + 2(.5) + 4(.7)$$

$$+ 6(.8) + 1(.9)]$$

$$= \frac{1}{25}(12.5) = 0.5.$$

5. In Exercise 4, C wins the votes of 7 voters at 0.8 or higher, A/B split the votes of 14 voters at 0.5 or lower, receiving 7 each. Since the 4 voters at 0.7 are closer to C than A/B, C wins a total of 7 + 4 = 11 votes, which is more than 1/3 of the total vote, despite the fact that less than 1/3 of the voters (7) lie at C's position or to his right.

6. Assume that A takes a position such that 1/3 of the voters lie to his left, B a position such that 1/3 of the voters lie to his right. By the reasoning given in the text, B can be beaten by a third candidate C who takes a position just to his right and thereby captures 1/3 of the votes. (A position just to B's left would give C fewer than 1/3 of the votes, since C would split the middle 1/3 with A.) But A now receives not only the 1/3 votes to his left but also splits the 1/3 in the center with B, thereby capturing more than 1/3 of the votes to

C's 1/3 (and B's less than 1/3). Now suppose that C takes a position somewhere in the middle 1/3 not adjacent and just to the right of B. Then he would receive less than 1/3 of the votes, because both A and B would win some votes in the middle 1/3. Thus, there is no position that a third candidate C can take that will guarantee him more than 1/3 of the votes, given A and B take positions to whose left and right, respectively, 1/3 of the voters lie. (For a more rigorous formulation and analysis of this question, see Theorem 1 in the Appendix.)

8. Such a joint strategy does not exist. By the reasoning given in the text, a third candidate C can always displace B, and a fourth candidate D can always displace A (except for ties). Thus, there exists no joint strategy of the two original candidates that makes it impossible for a third and fourth candidate to defeat both of the original two. (See Theorem 2 in the Appendix.)

9. Yes. M would receive votes from 4 voters at 0.6, 2 1/2 votes (!) from voters at 0.5, and 1 1/2 votes (!) from voters at 0.7, giving him a total of 8 votes, which is less than 1/3 (32 percent) of the total vote. C would receive 6 votes (all at his position or to his right), and L would receive the remaining 11 votes and win.

10. Yes. At the median 0.5, M would receive all the votes from the 5 voters at Md plus split the 4 votes at 0.4 with L and the 4 votes at 0.6 with C, giving him a total of 5 + 2 + 2 = 9 votes to 8 each for L and M.

11. At the median 0.5, M would receive 2 + 1 1/2 = 3 1/2 votes; at 0.4, he would receive 3 + 2 = 5 votes. Similarly, at any position between 0.3 and 0.5, M would also receive 5 votes; but at any position between 0.5 and 0.7 he would receive only 2 votes.

12. At the median 0.5, M would receive (atarting at 0.4) 4 + 5 + 2 = 11 votes. At any position 0.5 < m < 0.6, M

would receive (starting at 0.4) 4 + 5 + 4 = 13 votes, whereas at any position 0.4 < m < 0.5, M would receive (starting at 0.4) 4 + 5 = 9 votes. Thus, a position to the right of Md is optimal, given an extreme liberal and a moderate conservative.

13.

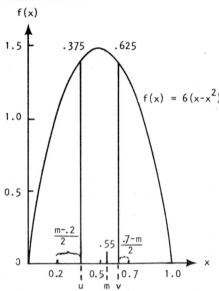

At any point m between x = 0.2 and x = 0.7, voters will vote for M in the interval u < x < v, where

$$u = .2 + \frac{m - .2}{2} = \frac{m + .2}{2};$$

$$v = .7 - \frac{.7 - m}{2} = \frac{m + .7}{2}.$$

The area under the curve in this interval is

$$A = \int_u^v f(x)\,dx,$$

which is at an extreme point when

$$\frac{dA}{dm} = f(v)\frac{dv}{dm} - f(u)\frac{du}{dm} = 0,$$

or

$$6[(v-v^2)(\tfrac{1}{2}) - (u-u^2)(\tfrac{1}{2})] = 0,$$

$$3[(v-u) + (u^2-v^2)] = 0,$$

$$3(u-v)(u+v-1) = 0 .$$

Substituting the expressions for u and v,

$$3(-.25)(\frac{2m-1.1}{2}) = 0,$$

$$m = .55.$$

Since $\dfrac{d^2A}{dm^2} = -\dfrac{3}{4}$, the extreme point is a maximum, and the number of voters is therefore maximized, when M's position is at m = 0.55. The area covered is in the center of the distribution between u = 0.375 and v = 0.625.

14. Either a position at the mode at 0.8 or at 0.6.

15. Wishy-washy, 3 votes; true, 9 votes; fuzzy, 15 votes.

18. No; x'y' would defeat all other platforms.

19. No; xy would defeat all other platforms.

20. No; x'y' would defeat all other platforms. Majorities, however, would be cyclical if platforms xy and x'y' were interchanged in the preference rankings of all three voters.

21. A complete interchange of two platforms simply involves a relabeling: what was platform P now becomes platform P', and vice versa. Since the underlying structure of preferences does not change, but only the labeling, cyclical majorities are unaffected.